Economics and Development Studies

Development studies textbooks and courses have sometimes tended to avoid significant economic content. However, without an understanding of the economic aspects of international development many of the more complex issues cannot be fully comprehended. *Economics and Development Studies* makes the economic dimension of discourse around controversial issues in international development accessible to second- and third-year undergraduate students working towards degrees in development studies.

Following an introductory chapter outlining the connections between development economics and development studies, this book consists of eight substantive chapters dealing with the nature of development economics, economic growth and structural change, economic growth and developing countries, economic growth and economic development since 1960, the global economy and the third world, developing countries and international trade, economics and development policy, and poverty, equality and development economists, with a tenth concluding chapter.

This book synthesises existing development economics literature in order to identify the salient issues and controversies and make them accessible and understandable. The concern is to distinguish differences within the economics profession, and between economists and non-economists, so that the reader can make informed judgements about the sources of these differences, and about their impact on policy analysis and policy advice. The book features explanatory text boxes, tables and diagrams, suggestions for further reading, and a listing of the economic concepts used in the chapters.

Michael Tribe is a development economist whose principal overseas research, consultancy and teaching experience has been in sub-Saharan Africa, in Uganda and Ghana in particular, over four decades. He has been the Honorary Secretary of the UK and Ireland Development Studies Association since 2000.

Frederick Nixson is Emeritus Professor of Development Economics at the University of Manchester. His research interests span macroeconomic policy, aid, industrialisation strategies and experiences, privatisation, the Asian transitional economies and poverty.

Andy Sumner is a fellow of the Vulnerability and Poverty Research Team at the Institute of Development Studies. He is a cross-disciplinary economist with primary foci of interest in child poverty and wellbeing; poverty indicators, concepts, methods, approaches; and the politics of policy processes.

Routledge Perspectives on Development

Series Editor: Professor Tony Binns, *University of Otago*

The *Perspectives on Development* series will provide an invaluable, up-to-date and refreshing approach to key development issues for academics and students working in the field of development, in disciplines such as anthropology, economics, geography, international relations, politics and sociology. The series will also be of particular interest to those working in interdisciplinary fields, such as area studies (African, Asian and Latin American studies), development studies, rural and urban studies, travel and tourism.

If you would like to submit a book proposal for the series, please contact Tony Binns on j.a.binns@geography.otago.ac.nz

Food and Development
E.M. Young

Natural Resource Extraction
Roy Maconachie and Dr Gavin M. Hilson

Health & Development
Hazel Barrett

Theories and Practices of Development,
2nd edition
Katie Willis

Economics and Development Studies

Michael Tribe, Frederick Nixson
and Andy Sumner

Routledge
Taylor & Francis Group

LONDON AND NEW YORK

First published 2010
by Routledge
2 Park Square, Milton Park, Abingdon, Oxon, OX14 4RN

Simultaneously published in the USA and Canada
by Routledge
52 Vanderbilt Avenue, New York, NY 10017

Routledge is an imprint of the Taylor and Francis Group, an informa business

© 2010 Michael Tribe, Frederick Nixon and Andy Sumner

Typeset in Times New Roman and Franklin Gothic
by Keystroke, Tettenhall, Wolverhampton, UK

All rights reserved. No part of this book may be reprinted or
reproduced or utilised in any form or by any electronic,
mechanical, or other means, now known or hereafter invented,
including photocopying and recording, or in any information
storage or retrieval system, without permission in
writing from the publishers.

Notice:
Product or corporate names may be trademarks or registered
trademarks, and are used only for identification and explanation
without intent to infringe.

British Library Cataloguing in Publication Data
A catalogue record for this book is available from the British Library

Library of Congress Cataloguing in Publication Data
Tribe, Michael A., 1943–
Economics and development studies /
Michael Tribe, Frederick Nixon, and Andy Sumner.
p. cm.
ISBN 978-0-415-45039-3 (hardback : alk. paper) —
ISBN 978-0-415-45038-6 (pbk.) 1. Economic development.
2. Economic development—Research. I. Nixon, Frederick.
II. Sumner, Andrew, 1973- III. Title.
HD77.T77 2010
338.9—dc22
2010002733

ISBN 13: 978–0–415–45038–6 (pbk)
ISBN 13: 978–0–415–45039–3(hbk)

Contents

List of figures

List of tables

List of boxes

List of acronyms

α	Incremental capital–output ratio (ICOR) – Greek lower case 'a'
ACOR	Average capital–output ratio
ACP	African, Caribbean and Pacific
BRIC	Brazil, Russia, India and China
BRICS	Brazil, Russia, India, China and South Africa
C	Aggregate Consumption
CA	Capability Approach
CAP	Common Agricultural Policy
CBR	Crude birth rate
CDR	Crude death rate
CMEA	Council for Mutual Economic Assistance (or COMECON)
CPIA	Country Policy and Institutional Assessment
Δ	Symbol meaning 'change in' – Greek capital 'D'
DDA	Doha Development Agenda
DFI	Direct foreign investment
DFID	Department for International Development
EMU	European Monetary Union
EOI	Export orientation industrialisation
EPAs	Economic Partnership Agreements (EU)
EPZ	Export processing zone
ERP	Effective rate of protection
EU	European Union
FDI	Foreign direct investment (can also be used as DFI – direct foreign investment)
GATS	General Agreement on Trade in Services
GATT	General Agreement on Tariffs and Trade
GDI	Gender Development Index
GDP	Gross domestic product

GEM	Gender Empowerment Measure
GFCF	Gross fixed capital formation
GM	Genetically modified
GNI	Gross national income
GNP	Gross national product
HDI	Human Development Index
HDR	Human Development Report (UNDP publication)
HIPC	Heavily Indebted Poor Countries
HPI	Human Poverty Index
I	Investment
IBRD	International Bank for Reconstruction and Development (part of the World Bank – WB)
ICOR	Incremental capital–output ratio
ICT	Information and communication technology
IDA	International Development Association (part of the World Bank – WB)
IDD	International Development Department (University of Birmingham)
IFIs	International financial institutions (i.e. the World Bank and the IMF)
ILO	International Labour Office
IMF	International Monetary Fund
IOM	International Organization for Migration
IPE	International Political Economy
ISI	Import substitution industrialisation
ISIC	International Standard Industrial Classification
K	National capital stock
LDCs	Less developed countries
MAI	Multilateral Agreement on Investment
MDGs	Millennium Development Goals
MDRI	Multilateral Debt Relief Initiative
MFA	Multi Fibre Agreement
MFN	Most favoured nation
MP	Marginal Product
MTEF	Medium Term Expenditure Framework
MVA	Manufacturing value added (contribution of manufacturing to national product)
NFCF	Net fixed capital formation
NGOs	Non-governmental organisations
NIEs	Newly industrialising economies
NIEO	New International Economic Order

OECD	Organisation for Economic Co-operation and Development
PPA	Participatory Poverty Assessment
PPP	Purchasing power parity
PRGF	Poverty Reduction and Growth Facility (of the IMF/World Bank)
PRSP	Poverty Reduction Strategy Paper
RBA	Rights-based approach
RMSM	Revised Minimum Standard Model (World Bank macroeconomic projection model)
RTA	Regional trading area (or regional trading agreement)
S	National savings
SITC	Standard International Trade Classification
SMEs	Small and medium-sized enterprises
TFP	Total factor productivity
TNC	Transnational corporation
TRIPs	Trade-related aspects of intellectual property rights
UDHR	Universal Declaration of Human Rights
UK	United Kingdom
UN	United Nations
UNCTAD	United Nations Conference on Trade and Development
UNDP	United Nations Development Programme
UNESCO	United Nations Educational, Scientific and Cultural Organization
UNFCCC	United Nations Framework Convention on Climate Change
UNIDO	United Nations Industrial Development Organization
UNMP	United Nations Millennium Project
UNRISD	United Nations Research Institute on Social Development
UN SNA	United Nations System of National Accounts
US$	United States dollars (where followed by a year – e.g. US$ 2000 – this refers to the value of dollars at constant prices for the year stated)
USA	United States of America
WB	World Bank
WC	Washington Consensus
WDI	World Development Indicators (World Bank publication)
WFP	World Food Programme
WIDER	World Institute for Development Economics Research (Helsinki – United Nations University)

WIID	World Income Inequality Database
WTO	World Trade Organization (previously GATT – General Agreement on Tariffs and Trade)
Y	National Income
Y_{EXP}	National Expenditure
Y_{DISP}	National Disposable Income

Foreword: this book and the authors

The origins of this book lie in two specific areas. First is the fact that, over a long period, observation has suggested that 'non-economic' members of the development studies community have tended to treat 'economics' as a difficult discipline in terms of both being intellectually demanding and being non-cooperative within a multi-disciplinary subject area. Many mainstream economists (and also some development economists) have tended to regard development studies as lacking the degree of rigour required of an intellectual discipline (particularly in terms of having an insufficiently robust body of theory and behavioural models). This has led to a mutual lack of sympathy between economics and development studies.

Second is a similar mutual lack of sympathy within the economics profession between development economists and 'mainstream' economists. The first issue is clearly understandable in justifying the efforts of the authors and the publisher in bringing this book into existence – an attempt to 'bridge the gap'. The second has been more difficult to explain as we drafted the chapters and discussed progress with the series editor.

Development studies is self-defined as a multi-disciplinary subject area. If students (and established professionals) within development studies regard the economic dimension as being peripheral to their understanding of international development, then this understanding will be only partial, due to the exclusion of important factors. Equally, if students and established professionals within economics perceive development studies as lacking the 'rigour' of economics (and particularly of its more quantitative and theoretical branches) and so do not regard communication as being 'productive', then their understanding of international development will also be partial. We have perfect conditions for a dialogue of the deaf.

The order of the authors' names on the cover of this book reflects our respective contributions. All three have discussed the overall approach and the contents. Michael Tribe has been principally responsible for drafting Chapters 2, 4, 5, 7 and 8. Frederick Nixson has been principally responsible for drafting Chapters 3 and 6. Andy Sumner has been principally responsible for drafting Chapter 9. Chapters 1 and 10 have been drafted on a more collective basis.

Michael Tribe is a development economist who started his academic career lecturing in the Department of Economics in what is now Makerere University, Kampala, Uganda in 1967, leaving at the end of 1971. Since then he has researched and lectured mainly in the Universities of Glasgow, Strathclyde and Bradford. Over the period mid-1982 to mid-1984 he was at the University of Cape Coast in Ghana. He has also undertaken short-term academic and consultancy assignments in Albania, Ghana, Kenya, Sierra Leone, Turkey and Uganda. His main areas of research and teaching interest have been development economics, industrial development, and project/policy analysis. Now semi-retired, he has been the Honorary Secretary of the UK/Ireland Development Studies Association since mid-2000, is an Honorary Visiting Senior Research Fellow in the Department for Development and Economic Studies, University of Bradford, and is an Honorary Lecturer in the Department of Economics, University of Strathclyde.

Frederick Nixson is Emeritus Professor of Development Economics at the University of Manchester. Before coming to Manchester he lectured in economics at Makerere University, Kampala, and has since lectured in a number of other countries, including Jamaica, Mongolia and Vietnam. His main research interests include industrialisation strategies and policies, trade issues, macroeconomic policy and the political economy of development. He has worked for the Asian Development Bank, World Bank, United Nations Development Programme and the Commonwealth Secretariat, in a variety of countries including the Yemen, India, Fiji, Solomon Islands, Ghana, Lesotho and North Korea. He is the author, co-author and co-editor of nineteen books.

Andy Sumner is a fellow of the Vulnerability and Poverty Research Team at the Institute of Development Studies. He is a cross-disciplinary economist. His primary foci of interest are: child poverty and wellbeing; poverty indicators, concepts, methods, approaches; and the politics of policy processes. He is Head of Graduate Programmes at IDS, and is a Council member of the European Association of Development Institutes and of the UK/Ireland Development Studies Association. His work to date has focused on Ethiopia, Tanzania, Uganda, India, Indonesia and Vietnam. He has conducted work for DFID, IFAD, Save the Children, UNICEF, UNDP and the World Bank Institute. Andy is the co-author of *International Development Studies: Theory and Methods in Research and Practice* (Sage, 2008), *After 2015: Development Policy at a Crossroads* (2009, Palgrave Macmillan) and *Children, Knowledge and Policy* (2010, forthcoming, Policy Press).

1 ▸ Development economics and development studies

1.1 Introduction

We are more aware of global poverty and inequality at the present time than at any previous time in history. Through the media, tourism and the campaigning activities of pressure groups and non-governmental organisations (NGOs), people in the developed market economies of Europe, North America, Japan and Australasia, both directly and indirectly, have an awareness of poverty and deprivation in low income economies that even a generation ago was inconceivable.

This explosion of information about developing countries and the experience of many people in living, working and travelling in these countries, is not always matched, however, by a better understanding of the reasons why poverty and deprivation persist, why new problems appear (the impact of climate change for example) and why economic growth appears to be so difficult to achieve and sustain over long periods of time (Nixson, 2002). There would be general agreement that in order to understand better the problems of economic and social development, some training in a social science (economics, sociology, social anthropology, political science or human geography) is needed.

There is little point in trying to discover 'who came first' among the social sciences in the study of development. Economics certainly has a long tradition. The famous economist W. Arthur Lewis was teaching development economics at the University of Manchester as early as 1950 (Leeson and Nixson, 2004), and increasing decolonisation in the

1950s and 1960s undoubtedly stimulated the study of development. But by the end of the 1970s, with an overwhelmingly disappointing record of economic development 'on the ground' and with political and social unrest widespread, many felt that a single disciplinary approach was not adequate, and that history, politics, sociology and geography at the very least had to be studied along with economics. This in turn led to the birth of development studies as an explicitly multi-disciplinary study of the development process in the 1960s (Sumner and Tribe, 2008a: Chapter 2).

The relationship between development economics and development studies has been a controversial one (Kanbur, 2002). Economics has been accused of imperialist tendencies not only within development studies but also across all the social sciences (Fine, 2002), and within development studies there is a perception of a tension between economists and non-economists (Harriss, 2002). Many of these controversies may appear impenetrable to the non-specialist, and also to those in the early stages of developing their interest in the study of international development. This book aims to help non-economists pick their way through some of the controversies, and to explain the very real contribution which economics (and development economics in particular) can make to the holistic multi-disciplinary approach of development studies. Briefly, the aims of this book are to:

a) Explain some fundamental economic dimensions of international development which are of the utmost significance to growth and poverty reduction in developing countries in the early twenty-first century;
b) Establish the role of economics in the understanding of international development in a cross-disciplinary (or multi-disciplinary) way;
c) Outline and explain some of the controversies within the economics profession which affect the study and analysis of the economies of developing countries.

It is not our intention to produce a textbook which competes with established (and new) books focusing comprehensively on the entire range of issues which fall conventionally into the area of development economics (or of the economics of development). There are several such textbooks which are readily available, and which are referred to at appropriate places in this book (see section 1.3 on Further Reading and Sources). The objective has been to provide a text outlining the economic dimension of international development studies, recognising that a) international development studies is a multidisciplinary subject area and that this economic dimension interfaces with other disciplines, and b) that for non-economists to be persuaded that this economic dimension is key to the understanding of international development, the presentation has to be accessible. The target readership of this book is best described as being at a level equivalent to a second- or third-year undergraduate studies degree relating to international development, but not involving specialist economic study.

It will become clear in later chapters that there is considerable controversy within the economics profession about appropriate analytical frameworks, methodology and methods/techniques. These controversies amount to clashes between alternative paradigms or *weltanschauung* (i.e. alternative ideological perspectives or 'views of the world'). While the differences within the economics profession apply on a worldwide basis, they have been particularly significant for developing countries subject to policy-conditionality based on the current dominant economics paradigm (which is 'neo-classical' with 'neo-liberal' overtones). These issues are discussed more fully in Chapters 2, 7 and 8.

Another area of controversy is between economists and non-economists in the context of analysis (i.e. intellectual understanding) of and policy prescription for developing countries. Mainstream economists have tended to criticise development studies specialists as lacking rigour, while development studies specialists have tended to criticise economists for an over-emphasis on measurable variables and on quantitative analysis together with an under-emphasis on qualitative analysis and historical/institutional perspective. The criticisms made by mainstream economists have been directed at both economists and non-economists working in Development Studies. Paul Krugman (a Nobel Laureate) has been particularly critical of Development Economics (Krugman, 1997), although his views are shared by many mainstream economists. Blow by blow accounts of the evolution of economic thinking and approaches to the study of developing countries' economies and of economic policy prescription are provided by Toye (1987, 1993 and 2003). Further discussion, from an economics perspective, of some of these issues may be found in Nixson (2006) and in Tribe and Sumner (2006).

It is possible to trace many of the concerns of development economics back to the earliest contributions of the nascent discipline of economics. Eighteenth-century writers such as Quesnay (1973), Mandeville (1970) and Smith (1974), for example, were concerned about the factors affecting economic growth, productivity growth, trade and income distribution. The historical and methodological tradition can also be traced in the nineteenth century through Marxian political economy (Marx, 1999), and the contributions of Malthus (1970) Ricardo (1953), and Schumpeter (1961, 1987).

Development studies is a younger field of academic endeavour, essentially cross- or multi-disciplinary in nature. Its evolution has been intertwined with that of development economics, with both developing as distinctive areas of enquiry after the Second World War. Development economics essentially emerged as an explicit 'labelled' area of study in the 1950s and development studies (again as a specifically labelled area of study) followed a little later. Several highly regarded international journals were established in the 1950s through to the early 1970s, including *Economic Development and Cultural Change* (1952), the *Journal of Development Studies* (1965), *Development and Change* (1970) and *World Development* (1973) (Sumner and Tribe, 2008a: Chapter 2).

The stimuli for the birth of development economics and development studies were very much framed around the decolonisation process (Sylvester, 1999). First, a push which was external to the developing countries – the 1949 Truman Declaration of 'a bold new programme . . . [to] make the benefits of *industrial progress* [emphasis added here]. . . available for the improvement and growth of under-developed areas' (cited in Esteva, 1992: 6) and the related Marshall Plan style of thinking that was influential in the 1950s and 1960s. Second, an interrelated, internal push (from within the developing countries) as newly independent states sought prescriptions for an 'economic catch-up' with industrialised nations (Shaw, 2004).

Development economics has always had a major concern for the issue of poverty reduction, and there has not been dependence upon purely economic (or income) approaches to poverty analysis in terms of both goals and policy (see Chapter 9 of this book) over the years. This involves less emphasis on income per capita and income poverty and more on multi-dimensional poverty exemplified in the United Nations Development Programme's Human Development Index and Human Poverty Index (UNDP, 2008) and in the UN international development targets for 2015, otherwise known as the Millennium Development Goals (United Nations, 2009a; Sumner and Tiwari, 2009a and b).

1.2 Development economics and development studies

There has been a tendency for some development studies-oriented textbooks to exclude the economics dimension, and there is evidence that some students regard economics as a 'hard' subject by comparison with other elements of development studies (such as sociology, anthropology, political science or geography). We feel that this tendency (and other evidence of 'economics-aversion') is very unfortunate – but not simply because we are economists. One aspect of the criticism of development studies from economists relates to the comparative absence of a rigorous theoretical basis, and another relates to the absence of quantifiable hypotheses and hypothesis testing. In many respects, the theoretical frameworks of, for example, social anthropology or of political science are of a different intellectual nature to those of economics, and they do not lend themselves to quantification or to an appeal to mathematical formulations. However, where issues in development studies lend themselves at least in part to statistical analysis, such analysis should not be avoided.

An example of the 'omission' of the economic, or statistical, dimension is provided by the otherwise excellent book by Allen and Thomas (2000) on poverty. While economists sometimes only consider the 'income' or economic dimension of poverty (which is itself a failing), for non-economists to ignore the economic dimension is just as much of a failing. The absence of Gini-coefficients and Lorenz curves, for example, from the Hewitt and Thomas book is a serious shortcoming. Our Chapter 9 attempts to achieve a balance between the contributions of a range of disciplines to discussion of poverty and poverty reduction issues.

Development studies is necessarily a cross-disciplinary subject. That is to say that the insights of a range of disciplines (or subject areas) need to be taken together in order to obtain a holistic view of international development. Cross-disciplinarity is a generic term which refers to the fact that more than one discipline is used in the type of study which is relevant to international development – and to development studies. Multi-disciplinarity refers to the fact that different individual disciplines are used collectively in development studies – while these disciplines retain their individual characteristics. Inter-disciplinarity refers to situations where the boundaries between disciplines (or subject areas) become blurred. For a fuller discussion of these issues see Sumner and Tribe (2008a: Chapter 3). The implication of this is that individuals working in the development studies area have a) to have some familiarity with more than one discipline or subject area, and b) to be able to work closely with other individuals who are specialists in other disciplines. Economics is but one of the disciplines which throws significant light on important issues in international development. A common misconception is that development studies is a subject area within the 'social sciences'. However, a little thought would make it clear that many of the major issues within development studies involve matters which require inputs from disciplines and subject areas within the natural or physical sciences – for example, areas such as agricultural production and environmental degradation. The cross-disciplinarity approach therefore transcends the social sciences.

Development studies textbooks and courses have sometimes avoided significant economic content. However, without an understanding of the economic aspects of international development many of the more complex issues cannot be fully comprehended. This book will make the economic dimension of discourse around controversial issues in international development accessible to second- and third-year undergraduate students working towards degrees in Development Studies. In addition, it is hoped that it will provide background reading for sixth formers, other undergraduate and postgraduate students and the informed general reader.

Features in this book include 'boxes' focusing on particular issues of interest, chapter by chapter listings of key economic concepts used in the text as well as suggestions for further reading, guides to relevant websites, and suggested questions for discussion.

1.3 Further reading and sources

Guidance about further reading and sources falls into three categories. The first concerns the subject matter of economics, the second concerns development economics, and the third concerns world wide web (internet) resources.

The first line of 'defence' when looking for explanation of economic concepts might be to reach for a basic economics textbook (of which there are many) and to search the subject index. Many such textbooks now have their own dedicated websites which can help considerably in searching for such explanations. However, an alternative approach is to use an economics dictionary, several of which are readily accessible. One of these is the *New Palgrave Dictionary of Economics* which has many short explanatory articles about specific economic issues (Durlauf and Blume, 2008). This dictionary also has a searchable website, and an article on development economics by Ray (2008). *The Economist* magazine has an excellent online dictionary of economic terms which is called *The Economics A–Z* (*The Economist*, 2009). The Economics Network (2009) is devoted to university-level economics teaching, and many explanatory resources are available online. Although the online encyclopaedia Wikipedia is not acceptable as an academic source, it does contain a large number of very informative articles about specific issues – including economic issues (Wikipedia, 2009). The reason that Wikipedia is not acceptable as an academic source is that it is not 'peer reviewed' (meaning that anybody can upload material onto the website without checking or oversight). However, it is of course possible to read *Wikipedia* articles and then to double-check other sources which are more academically acceptable. An accessible and brief source for clarification of many economic concepts is Dasgupta's *Economics: A Very Short Introduction* (2007), which has synergies with, but is not focused on, 'development'.

There are many development economics textbooks, several of which are regularly updated with new editions, and with their own dedicated websites. Two of the leading textbooks are Todaro and Smith's *Economic Development* (2008), and Thirlwall's *Growth and Development* (2006), both weighing in at about 800 pages. A new addition to this market is *Development Economics* by Clunies-Ross, Forsyth and Huq (2009). Each of these books has its own particular approach, and they aim to cover the entire range of issues within the concern of development economics. A textbook which has a more distinctive approach, without aiming to cover the entire range of subject matter, is *The Process of Economic Development* by Cypher and Dietz (2008). Routledge has a range of titles in a special development economics section within their current *Economics* list.

There are many internet websites which are relevant to international development, and to the economic dimension, and it is necessary to limit this introductory discussion. One very good starting point which is surprisingly little known is the UK Department for International Development's *Glossary* (DFID, 2009). This glossary covers not only a range of acronyms ranging from the well-known to the obscure, but also a large number of invaluable explanations of terminology in common use in the international development field. A very important source for information about current research in the development studies area is the *id21* website (id21, 2009). This website contains a wealth of brief summaries of recent publications and working papers together with weblinks which permit the downloading of fuller reports. Another website based at the Institute of Development Studies at the University of Sussex is ELDIS (ELDIS, 2009). This is essentially a 'portal' which gives access to an incredibly large number of sources of information about international development.

The websites mentioned in the previous paragraph are quite 'robust', meaning that they are 'long term' and that it will be possible to find them easily at or around the current location for a considerable time to come. However, the world wide web is constantly changing, with new names, changed names and changed locations sometimes making it difficult to find resources. This makes 'search engines' particularly useful instruments for literature searches. The most successful search engine in recent years has been Google (Google, 2009). While the basic search engine facility can be effectively used to find details of publications and of documentation, there are two specialist instruments which are of particular academic interest. The first of these is Google Scholar (2009), which allows searches for a wider range of journal articles than most similar instruments. The second is Google Books (2009) which allows the full text of an increasing number of books to be seen on the computer screen. In Google Books the text of some books is made available directly by the publisher, while in other cases Google have made their own arrangements (with permission from the publishers) to make the text visible digitally. Many universities now have their own software for 'online learning' (one of which is Blackboard) and their learning support services provide full explanations for their use.

1.4 The structure of this book

Following this introductory chapter, *Economics and Development Studies* consists of eight substantive chapters with a final chapter which contains some concluding remarks. Chapter 2 deals with the nature of development economics, an issue which can excite some controversy amongst economists. Chapter 3 discusses the nature of structural change in economic development, particularly focusing on the importance of industrialisation. Chapter 4 outlines some of the economic approaches to, and the significance of, economic growth. Chapter 5 presents some evidence for the economic growth and development of both developing and developed industrial countries over the period since the Second World War. Chapter 6 explores the phenomenon of globalisation, while Chapter 7 focuses on issues relating to trade and developing countries. Chapter 8 discusses a number of economic aspects of development policy, and Chapter 9 concentrates on the important topic of poverty. The concluding chapter rounds off the book, reflecting on some of the current and future issues affecting developing countries.

For a book of this length there are obviously many important issues which have been omitted or which have been dealt with only briefly. Wherever possible, we have tried to refer to other sources which address such issues more fully.

Suggested further reading

Clunies-Ross, A., Forsyth, D. and Huq, M. 2009. *Development Economics*. London: McGraw-Hill.

Cypher, J. and Dietz, J. 2008. *The Process of Economic Development* (3rd edn). London: Routledge.

Dasgupta, P. 2007. *Economics: A Very Short Introduction*. Oxford: Oxford University Press.

Thirlwall, A.P. 2006. *Growth and Development with Special Reference to Developing Countries* (8th edn). Houndmills, Basingstoke: Palgrave Macmillan.

Todaro, M. and Smith, S. 2008. *Economic Development* (10th edn). London: Pearson Addison Wesley.

2 The nature of development economics

2.1 Introduction

In Chapter 1 we have defined development studies as a multi-disciplinary (or cross-disciplinary) subject area rather than as a 'discipline' in itself. Economics has a long intellectual history as a discipline, comparable with other disciplines such as sociology, mathematics or chemistry. However, the study of economics itself involves a considerable amount of 'inter-disciplinarity' with, for example, technical/scientific issues being of particular significance in production economics, including the economics of agriculture, and in environmental economics (see for example Sumner and Tribe, 2008a: Chapter 1 and 2008b). This interaction with other disciplines is considerably greater in what is known as 'micro economics', which relate to – inter alia – production (the firm) and consumption (the household), than in 'macroeconomics', which relate to the economy as a whole, including national income analysis, which covers aggregate savings, investment and international trade and payments. We view economics as an integral part of a multi-disciplinary development studies, but retaining a distinctive role specific to its individual disciplinary characteristics within the broader multi-disciplinary framework.

In the Preface to the first edition of his invaluable collection of essays, *Global Political Economy*, Ravenhill (2005a: v) has some revealing remarks:

> In my experience, the question that students most frequently ask
> on enrolling in a course in international political economy is
> whether they need a background in economics to do well. My
> answer is always that IPE is a sub-field of international relations,
> that the course is in political science not in the discipline of
> economics: a background in economics is neither necessary nor
> sufficient for doing well in IPE.

The suggestion that IPE (International Political Economy) is a
sub-field of international relations within political science, and is
not in the discipline of economics essentially denies the existence
of multi-disciplinarity and asserts the non-violability or inflexibility of
disciplinary boundaries – a view which will be an anathema for most
students of development studies. However, not least, any attempt to
place 'political economy' within the discipline of political science
runs counter to centuries of intellectual endeavour by the economics
profession. The intention in this chapter is to place multi-disciplinarity
at the forefront of the discussion and to demonstrate how the
discipline of economics can contribute substantially to the study
of international development.

A widely held view in the economics profession is that economic
principles and theory are essentially universal, and that they are
equally applicable to advanced industrial and developing countries.
We need to address this view, and to distinguish between a) the
implied universal application of basic concepts, principles, methods
and theories within 'economics', and b) the adaptation and application
of economic concepts, principles, methods and theories to the varied
socio-economic characteristics of developing countries and regions
which are the focus of development studies.

The discussion of diverse views of the role of economics in the study
of developing countries has been the focus of many publications over
the years. One of the most recent is based on an international
conference hosted by the United Nations University's World Institute
for Development Economics Research (WIDER) in 2005 reviewing
the current state of development economics, the proceedings of which
have since been revised and published (Mavrotas and Shorrocks,
2007). The conference was very comprehensive in covering both
contemporary issues and long-running controversies. A recent
introduction to a collection of articles reviewing '50 years of
development economics' (Tribe and Sumner, 2006) is followed by a
call for a return to a 'classical' focus on accumulation, growth and

structural change by Nixson (2006). Nixson's article follows on from a substantial review of the role of economics in the cross-disciplinary study of development undertaken by Leeson (1988) and by Leeson and Nixson (1988 and 2004).

There is a view amongst some development economists that although the main body of 'economics' is broadly applicable to developing countries, the limited availability of economic statistics and the poor quality of many of the available statistics make it difficult to undertake rigorous evidence-based economic analysis (and particularly quantitative analysis) as effectively in developing countries as in advanced industrial countries. The pragmatic approach, which involves picking and choosing analytical approaches according to whether they help to understand the phenomena being studied (often in a somewhat eclectic way), is essentially the position taken by 'heterodox economists'.

Labelling in the economics profession can be idiosyncratic. The conventional wisdom tends to be 'neo-classical', but there is resistance to the relatively short-run nature of neo-classical analysis and to the often value-laden implications of its myopic application. Development economists who are antagonistic to a strict neo-classical, or 'mainstream', approach have been labelled 'structuralist' in the past, but more currently fit most easily into 'heterodox economics'. Another sub-group within the economics profession which aligns with heterodox economics is 'supply-side economics', espousing an interventionist approach rather than an unquestioning market-driven approach (Mearman, 2007; Lee, 2008).

Another important distinction is between the concept of 'development economics' as a division within the discipline of economics (comparable, for example, with regional, industrial, agricultural or health economics) with its own specific theories, methods and approaches and having a distinct specialisation, and the concept of the 'economics of developing countries' which involves the application of economic principles to developing countries. In this chapter, we refer to the application of economics to the 'big picture' as 'development economics' and to the more routine application to detailed analysis of development issues as the 'economics of developing countries'.

The remainder of this chapter will consist of a discussion of the relationship between economies and development studies/theory in section 2.2, further consideration of the meaning of the term 'economics of developing countries' in section 2.3, and a review of economic paradigms, 'schools of thought' and interaction between discrete disciplines and subjects in section 2.4. Section 2.5 provides a rounding off of the chapter with some overall conclusions.

2.2 Development economics and international development studies

Development studies embraces a body of principles and constructs which help to explain the socio-economic development of individual countries, global regions and the international system, and to make predictions about the future. It is concerned with the 'big picture': what are the driving forces behind the long-term evolution of human society?; why have some countries attained high standards of living while others have not?; why do some countries have flexible socio-economic systems which are amenable to change and others do not? Many of these questions are, by their very nature, multidisciplinary. No single discipline or subject can claim to have complete insight into the factors which explain 'development' or into the interaction between these factors (Sumner and Tribe, 2008a).

In the 1960s and 1970s, there was a considerable emphasis on broad strategical approaches to economic development. They included the 'Big Push' or 'Critical Minimum Effort' (Rosenstein-Rodan, 1943; Leibenstein, 1957: Chapter 8), Balanced vs Unbalanced Growth (Hirschman, 1958: Chapters 3 and 4; Streeten, 1959) and the Stages of Economic Growth (Rostow, 1960a and 1960b). The first two of these depended critically on the interaction of external economies and inter-sectoral linkages (including the relationship between agriculture, industry and economic infrastructure), while the third was based on broad historical interpretations of economic growth and development. In a set of lectures delivered in 1992, Krugman was critical of these broad approaches since he considered them too vague and insufficiently focused on evidence-based analysis (1997: 27–8). However, since the 1980s there has been considerably less emphasis on these strategical approaches by development economists. The evolution of thinking within development economics over the last forty years is perhaps most easily traced through browsing the contents of Meier's *Leading Issues* from the first edition (Meier, 1964) to the eighth edition (Meier and Rauch, 2005), the contents of

Todaro's *Economic Development* from the first edition (Todaro, 1977) to the tenth edition (Todaro and Smith, 2008) and the contents of Thirlwall's *Growth and Development* from the first edition (1972) to the eighth edition (2006). It is significant that although the first edition of Meier's edited collection was entitled *Leading Issues in Development Economics*, from the third edition the title was changed to *Leading Issues in Economic Development*. Influential development economics textbooks of the 1960s included Higgins' *Economic Development* (1968) and Kindleberger's *Economic Development* (1958), both of which have long since been replaced in undergraduate and postgraduate reading lists, although they are still of scholarly value in terms of the clarity of their exposition and as part of the intellectual development of the subject. Another very significant contributor to the literature on the analysis of economic development was Dudley Seers (1959 and 1962 for example).[1]

Some of the most influential theories of economic growth and development are based on quite simple economic concepts and principles. One of the most widely cited theoretical approaches to economic development was first published by Arthur Lewis in the mid-1950s (Lewis, 1954). Over the years, Lewis followed up his 1954 article on 'Economic Development with Unlimited Supplies of Labour' with further publications responding to reactions and making elaborations to the original article (refer to Ghosh, 2007 and to Tribe and Sumner, 2006). Willis (2005: 42) refers to the 'Lewis model' in the context of structural change, and particularly focuses on its power to explain the changing balance between urban (industrial or 'modern') and rural (agricultural or 'traditional') population. However, the Lewis model is principally, as emphasised by Lewis himself, a basis for the explanation of the process of capital accumulation through re-investment of profit (or surplus) generated in the modern sector on the basis of a wage rate maintained at a low level because of the plentiful ('unlimited') supply of cheap labour from the traditional sector. These economic dimensions of the Lewis model are therefore fundamentally within a 'theory of development' but if these are omitted from a 'development studies' dimension, the model loses much of its power. The Lewis model is very much in the intellectual tradition of classical political economy, although the conventional diagrammatic presentation in Figure 2.1a summarises some of its main features through the use of basic neo-classical economic concepts.

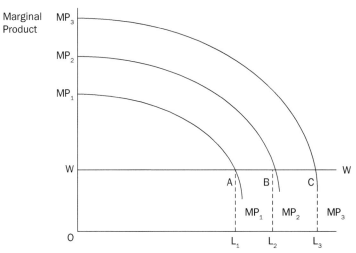

2.1a A diagrammatic presentation of the 'Lewis model'

In Figure 2.1a, the horizontal axis represents the level of employment in the modern sector (measured in numbers of workers) and the vertical axis represents the value of the marginal product associated with different levels of employment. The 'unlimited supply of labour' is represented by the horizontal line indicated at W, showing that at every level of employment the wage rate remains the same – the marginal wage rate (the addition to total labour costs due to the employment of one additional worker) is constant, meaning that the average wage rate is equal to the marginal wage rate. The total wage cost is shown by the area below the line W at any given level of employment (employment multiplied by the average wage rate – e.g. $OWAL_1$). The 'production functions' are the lines $MP_1\ MP_1$, $MP_2\ MP_2$, and $MP_3\ MP_3$. Individually each production function shows that as additional labour is added to a fixed capital stock, the level of total output rises but at a decreasing rate, indicating diminishing marginal returns. Investment increases the size of the capital stock and is represented by the shift from one production function to another, higher one – for example, from $MP_1\ MP_1$ to $MP_2\ MP_2$ showing that for the same amount of labour a higher level of output can be achieved.

The vertical axis in Figure 2.1a shows the marginal product at different levels of employment. At the level of employment L_1, the value of total production is shown by the area OMP_1AL_1. This means that the difference between the value of production and the cost of employing the labour force is 'profit' or 'surplus value' received by the owners of the enterprises in the modern sector – this is shown by the area MP_1AW in the figure. From an economic point of view, the significant issue is that in Lewis' model the capitalist owners re-invest this profit, increasing the size of the capital stock in the modern sector, increasing the demand for labour from L_1 to L_2, and shifting the production function to $MP_2\ MP_2$. This has the effect of increasing profits to MP_2BW, which are again re-invested. This results in a further increase in the demand for labour, a shift of the production function to $MP_3\ MP_3$, and an increase in the level of profits to MP_3CW.

If the increase in the demand for labour at each of these stages were to lead to higher levels of wages – i.e. if the unlimited supply of labour were to 'run out' – this would be shown either by an upward shift of the labour supply curve WW to a higher level (from W_1W_1 to W_2W_2 in Figure 2.1b(i)), or by a rising curve in the line W_1W_1 indicating that after a certain point additional labour was only available at higher wages (shown in Figure 2.1b(ii)). It can easily be seen that when the 'unlimited supply of labour' 'runs out' it adversely affects the level of profits and the rate of re-investment in an expanded capital stock.

In his critical review of 'development economics', Krugman (1997) focused on Lewis' 'Economic Development with Unlimited Supplies of Labour' as being 'probably the most famous paper' in development economics, but suggests that 'it is hard to see exactly why'. He takes the view that 'during the years when high development theory flourished, the leading development economists failed to turn their intuitive insights into clear-cut models that could serve as the core of

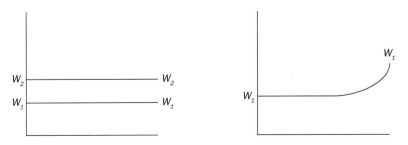

2.1b *Does unlimited labour 'run out'?*

an enduring discipline' (1997: 24). His overall intellectual position is that 'economic theory is essentially a collection of models', and that 'the influence of ideas which have not been embalmed in models soon decays . . . and this was the fate of high development theory' (1997: 27). Being renowned in the area of spatial economics (and, indeed, being a Nobel Laureate), Krugman's views clearly need to be taken seriously. However, others within the economics profession obviously take a different view because there is still a significant amount of emphasis given to Development Economics institutionally (within international bodies, research institutes and departments throughout the world), in terms of publications (such as books and specialised journals), and in university teaching programmes.

Lewis' theoretical construct is more complex than has been presented in the simplified version outlined in this chapter. However, in many respects the Lewis model simply formalises some of the basic relationships which have occurred in most developed countries as they passed through early 'stages' of industrialisation. Lewis did not summarise his articulation of the economic relationships in diagrams or equations. The Lewis model employs very basic economic (mainly microeconomic) concepts within discussion of the structural characteristics (at a macroeconomic level) of developing countries in order to arrive at an elegant and powerful theoretical framework which increases understanding of long-term economic development. Indeed, the basic ideas expressed within the model can be traced back to the classical political economists of the nineteenth century. A conference to celebrate the fiftieth anniversary of the publication of Lewis' original article was held in the University of Manchester, and the proceedings were published in the Manchester School (see for example Kirkpatrick and Barrientos, 2004). An interesting survey article by Ghosh (2007) explores the extent to which the terminology and concepts used by Lewis have been lost or misrepresented over the years. Extended discussion of the Lewis model in a development economics textbook context can be found in Ingham (1995: Chapter 4), Thirlwall (2006: Chapter 5) and Todaro and Smith (2008: Chapter 3). Ranis and Fei (1961) have an elaborate neo-classical restatement of what is essentially the Lewis model in a seminal article.

In his criticism of the Lewis model, Krugman (1997) marshals arguments which are characteristic of the views of 'mainstream' economists in criticising 'development economics' – or even in denying its existence.[2] 'Modern' economics has tended to become ever more quantitative and mathematical in expounding theory and in undertaking rigorous empirical analysis. Elsewhere in this chapter we have explained how both the quantity and the quality of economic statistics available for developing countries exert serious limitations on economic analysis – not least on its quantitative variety. However, a theory or a 'model' does not have to be expressed in algebraic terms in order to communicate logical structure and predictive capacity (two of the main attributes of a theory). In the remainder of this section, we have outlined areas of the study of 'development' by economists which are parts of a body of 'development economics'.

The first relates to internal domestic as well as to international development, and is concerned with forces leading to increased inequality – namely, the 'centre–periphery' approach. While this approach does not usually include any explicitly formal presentation, such as the diagrammatic depiction of the Lewis model, it has sufficient substance to be included here. Myrdal (1957) is probably the originator of this approach from the relatively early days of development economics, referring to 'spread' and 'backwash' effects within a 'mechanism of national and international inequality', and more recently Thirlwall has used the same concepts for the analysis of international trade (1983 and 2000). Kay (1989: Chapter 2) regards the centre–periphery approach as being central to the structuralist economists' theory of economic development. While this approach cannot necessarily explain initial inequality within the global economy, it can provide a robust basis for a process of cumulative causation leading to reinforcement and intensification of international inequality. Further development of the issues addressed by the centre–periphery approach in the analysis of international development occurs within 'dependency theory' and neo-Marxian schools of thought, which are in the traditions of political economy. Many of the ideas associated with these radical writings arose from Latin American experience and are well represented by Roxborough's influential book (1979).

The second example is an analytical framework describing the nature of the international trading relationship between developed industrial countries and developing countries – referred to as the Prebisch–Singer thesis. The original publications relating to the Prebisch–Singer thesis are Prebisch (1950) and Singer (1950). Essentially, the two authors reached the same conclusions independently. There have been many re-capitulations, critical reviews and defences of the basic propositions over the years and many of these are reviewed in a special issue of the *Journal of International Development* (Sapsford and Chen, 1999).

The Prebisch–Singer thesis is closely associated with the centre–periphery approach in the sense that it describes a process of cumulative inequality. The basic proposition of the original version of the thesis is that, due to the interaction between the market characteristics of developing countries and developed industrial countries, the terms of international trade between the two groups will – in the long term – tend to move against developing countries. The expression 'terms of trade' refers to the relative prices of imports and exports, and if the terms of trade 'move against' a country or group of countries, it means that their export prices are falling relative to their import prices. Put simply, if the terms of trade are deteriorating, it will be necessary to *export* larger volumes of goods and services in order to *import* an unchanged volume of goods and services. If one side of the trading relationship is suffering from declining terms of trade, the clear implication is that the other side is benefiting from improving terms of trade. Another way of looking at deteriorating (or declining) terms of trade is that they represent a transfer of income away from those with declining terms towards those (the other 'half' of the relationship) with improving terms of trade, and this was the principal point made by Singer (1950).

It does not take much imagination to extend the Prebisch–Singer thesis to a theory of exploitation of developing countries by developed industrial countries through the international trading system. However, many economists would not extend the argument that far. Instead, they would argue that it is in the nature of markets that some products experience falling prices and others experience rising prices due to changing patterns of demand and to changing technology and production systems. For these market-oriented economists, falling prices would be a signal that producers should move out of production of these commodities and find other, more profitable, commodities to produce and export. Essentially, this view regards the world as having a constantly changing economic system (see Box 2.1).

The original Prebisch–Singer thesis was formulated as an explanation for observed phenomena in international trade. There has been a considerable amount of controversy around empirical observation of

---

Box 2.1

The basic economics of the Prebisch–Singer thesis

The types of primary commodities considered by Prebisch and Singer included raw materials such as cotton, jute and sisal and beverages such as coffee, tea and cocoa. The argument set out by Prebisch and Singer has recently been extended to include 'simple' manufactures, including basic textiles and garments and routine assembly of electrical products.

The commodities are produced and exported in a competitive market. This means that there is comparatively easy entry and exit, the production technology does not involve significant obstacles to entry such as complex technology or management requirements and does not involve significantly high levels of investment for new entries. In these circumstances, the producers are 'price takers' – no matter how much or how little they export, any change which they individually make to supply has no effect on the world market price. The price elasticity of demand and the price elasticity of supply are both high.

Many of the products considered in these types of markets are not only produced in several countries, but they can also themselves easily be substituted – there is a high elasticity of substitution. This means that if the price of one commodity increases, it is easy for importing countries and for the consumers in these countries to switch to an alternative product. Some of these alternatives are synthetic substitutes. Examples of this include switching between different beverages and between different types of textiles.

The commodities being considered account for a small proportion of expenditure in the importing countries and have a low-income elasticity of demand – as incomes increase, the expenditure on these items tends to increase by a lower percentage than incomes.

To summarise – the markets for these commodities have high-price elasticities of demand and supply, high-price elasticities of substitution and low-income elasticities. The effect of this combination of economic characteristics is that the products have a 'weak' market position, leading to the effects which Prebisch and Singer identified.

---

the international terms of trade. Some economists argue that there is no significant evidence to support the 'theoretical' propositions of Prebisch and Singer – the disputes revolve around which particular groups of commodities, which countries, and which time periods should be included in the analysis. Different sets of data give different conclusions. Some of these disputes are summarised in the main development economics textbooks (good examples of this are Thirlwall, 2006: Chapter 5 and Todaro and Smith, 2008: Chapter 3), and further exploration of these issues will be found in Chapter 7 of this book.

It is significant that a mid-1990s World Bank publication concerned with Structural Adjustment experience in sub-Saharan Africa took the view that international aid was a positive feature because, for a number of countries, it had compensated for national income lost through the adverse movement of the international terms of trade (Husain, 1994: 7), although this is not usually considered to be the major objective of international aid.

A further development of this theme in more recent years has been the extension of the Prebisch–Singer framework to include the production of manufactures in developing countries and their export to developed industrial country markets. This approach distinguishes between 'routine' types of manufactures which are produced in and exported from developing countries, and higher technology products which are manufactured in and exported from developed industrial countries. Essentially, the point is that the export markets for routine manufactures share many characteristics with those for primary commodities – the markets are very competitive, with relatively high price elasticities of supply and demand, and with comparatively low income elasticities – so that, it is claimed, the deteriorating terms of trade observed for exporters of primary commodities also apply to exporters of routine manufactures (Sarkar and Singer, 1991). This issue is also considered in Chapters 3 and 6 of this book on structural change and on globalisation, with the publications of Sanjaya Lall and associates being particularly significant (for example Lall et al., 2006).

In the context of the economic theory relating to international trade, there have been other developments more recently. A major initiative is the 'North–South' approach, focusing on the mechanisms which provide an analytical basis for the 'uneven development' which is observed internationally. This approach to the understanding of the forces which explain patterns of international development very much builds on the concepts within the Prebisch–Singer thesis and within the centre–periphery approach, both of which relate to asymmetrical development tendencies. One of the main proponents of the North–South approach has been Krugman (1981), but further elaboration has been provided by Ocampo (1986) and by Dutt (1988),

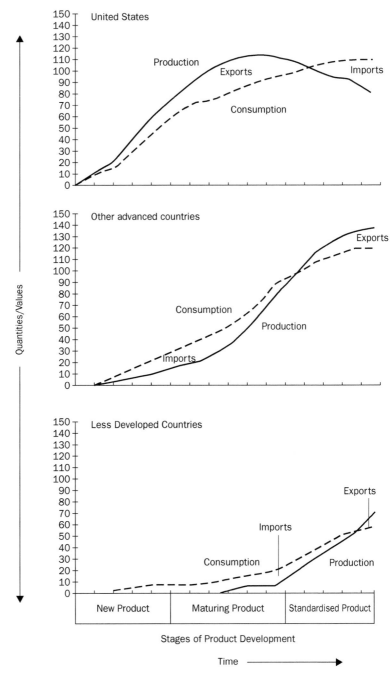

2.2 Vernon's original product cycle

Source: Vernon (1966: 199 and 1971: 449)

all working within the framework of 'new trade theory' as summarised more recently by Sen (2005).

A second example of a 'model' of international trade patterns in the context of developing countries is Vernon's product cycle, which first appeared in an article published in 1966 (Vernon, 1966) and which is presented diagrammatically in Figure 2.2. Vernon later published a reconsideration and partial retraction of the original arguments (Vernon, 1979). The basic idea of the product cycle is that new products tend to be conceptualised and developed in industrially advanced countries. New products often also involve new production technologies, with degrees of technical sophistication which need to be 'worked out' in the early stages of production. Eventually the production of these new products becomes 'routinised' and is transferred to countries with lower levels of industrial development (and with lower wage levels). The countries which originally developed the new products then increasingly import the 'routinised' product. The industrial sector in more advanced countries replaces production of the 'routinised' products with a further round of new products, retaining a degree of dynamism in this process. The fact that industrial research and development tends to be centralised in the more developed countries reinforces their market dominance. This approach fits well, in general, with a process of industrialisation in which the transnational corporations are major actors. While the detailed evidence relating to this model is open to endless wrangling, there is a certain attraction in its simplicity, and it contains a sufficiently significant germ of truth to be a useful concept to use in the analysis of global industrial development in the second half of the twentieth century (Williamson and Milner, 1991: 73–5).

Although it did not have any significant evidential or quantitative support in its original form, Vernon's model is intellectually, logically and intuitively attractive, and fits a process of international development which still continues. Here we have an example of an elegant theory which is associated with economic development, which contains the essence of a 'model' in Krugman's sense, and which produces refutable propositions. From Krugman's viewpoint the weakness of this model is that it does not have a quantitative or mathematical form. However, it is possible to appeal to evidence on the basis of which the predictions of the theory can be verified – if agreement were to be possible between opposing camps of disputing economists about the validity and relevance of evidence.

2.3 The economics of developing countries

What is the distinction, then, between 'development economics' and the 'economics of developing countries'? Economics is a very diverse area of study, and includes a large number of specialisms. Some specialist areas relate to particular subjects, such as health economics, where the technical and other specific aspects of the subject imply that individual economists need to develop a relevant stock of technical knowledge and experience in order to work effectively in the area. A transport economist, for example, would experience great difficulty working on detailed aspects of health economics. Other specialist areas relate to familiarity with particular countries or regions of the world. In order to work effectively on the economies of these countries and regions, it is necessary to build up specialised local knowledge. In some cases – such as working on health issues in sub-Saharan Africa – it may be necessary for specialist health economists to work with health specialists (for technical aspects), sociologists and development economists with local knowledge in multidisciplinary teams. However, there are many fundamental concepts and theories which are shared by all branches of economics, but without implying the adoption of the full neo-classical 'toolkit', and certainly not implying the adoption of a neo-liberal or 'market fundamentalist' ideology (see Chapter 8 for further elaboration of this issue).

There are also different analytical methods with which particular economists have familiarity, and which may be brought to bear on a subject area (e.g. health, transport or agriculture) or on a specific country in collaboration with other specialist economists. Finally, there are different 'perspectives' or paradigms – in part related to intellectual or ideological positions – which provide a further fragmentation of the economics profession. There is imperfect substitutability between different types of economists just as there is imperfect substitutability between different branches of the medical profession – in other words, there are 'horses for courses'. One of the implications of this fragmentation of the economics profession is that in addition to the need for cross-disciplinary teamwork between specialists from different disciplines, there is often a need for teamwork between different types of economists in addressing a particular analytical or research problem.

How is this discussion relevant to the theme of this book and of this chapter? If there is indeed a valid specialist area of 'Development Economics', this implies that the subject area has a degree of exclusivity which applies to both theoretical and empirical economic studies within developing countries. However, the 'Economics of Developing Countries' perhaps implies much less exclusivity, so that a body of universal economic concepts, principles, methods and theory are applied to developing countries – and these concepts, principles, methods and theory are also applicable to developed industrial countries with radically different structural characteristics. This approach invites a questioning of the meaning of 'structural characteristics' in the economic context. Box 2.2 and the following discussion aim to provide an answer to this issue.

The significance of the differences between the economic characteristics of developing and developed industrial countries is that these characteristics determine economic behaviour, at both the economy-wide *macroeconomic* level and the firm or household *microeconomic* level. Although the economic principles and concepts employed in analysing various aspects of the economy – including growth and development – will be essentially the same for all types of economies, the interactions between these economic characteristics in various economic structures will be different. For example, predictions of future events in one type of economy cannot be based on experience of economies with different structural characteristics, and predictions of the economic responses to policy-based changes in significant variables (such as the foreign exchange rate, interest rates or tax rates) in one type of economy cannot be inferred to apply to economies with different structural characteristics. This is the essence of the argument of 'structuralist' economists, and also the basis for much of 'heterodox economics' – which have been referred to earlier in this chapter.

Earlier in this chapter reference was also made to two articles by Seers published in 1959 and 1962, both of which were reaching towards a 'structuralist' (as opposed to a 'neo-classical') approach to the macroeconomic analysis of economies in Latin America. Over the years this approach evolved and has even been accepted by economists who would be regarded as within the neo-classical school: a notable contribution was by Polak (1997) extending over three decades and amounting to an 'open-economy macroeconomics' approach. This

Box 2.2 A stylised dichotomous comparison of the structural characteristics of economies[a]

Characteristic	'Developing countries'	'Developed industrial countries'
National income/value added/production	High proportion of national income originates in agriculture and other primary production, and a low proportion in manufacturing. Relatively small proportion of large-scale enterprises. Comparatively high proportion of small-scale enterprises in all sectors of the economy.	Very small proportion of national income originates in agriculture and other primary production, and a comparatively high proportion in manufacturing. Relatively high proportion of large-scale multi-branch enterprises in all sectors of the economy.
Government taxation and revenue	Limited tax base: personal income tax restricted due to low incomes and high rate of self-employment; company profits tax restricted due to large number of small enterprises and accounting compliance – value added tax restricted for the same reasons. Relatively high costs of collection.	Well-developed tax base and considerable 'self-administration' for personal and business taxes. Wide range of tax sources and relatively low costs of collection.
Financial infrastructure	Low proportion of the population hold bank accounts, with many transactions based on cash and other non-electronic means. Mix of locally and internationally owned banks, with frequent problems monitoring and enforcing banking regulations for locally owned institutions. Significant 'non-performing assets' (bad debts) problem for locally owned institutions.	High proportion of transactions undertaken electronically. Banking sector highly globalised. Banking regulations and supervision highly developed.[b]
Employment and social protection	High proportion of self-employment among working-age population. Significant proportion of family and casual labour with informal or non-	High proportion of wage/salary employment among working-age population. Most employees have contestable contract of employment, and significant

	existent 'contract' of employment and low employment/safety protection. Absence of unemployment benefit schemes and lack of formal 'employment guidance' schemes.	employment and safety protection. Unemployment benefit and 'employment guidance' widely available. Established social security benefit schemes.
Education system	Primary schooling now approaching a universal coverage, but with issues over quality. Secondary education has increased considerably in the last fifty years. Further and higher education coverage still very limited. Gender issues arise in some countries. Relatively low (but increasing) literacy and numeracy rates.	Primary and secondary schooling now essentially universal. Proportion of the population with further and higher education qualifications is now significant. Gender equity has increased significantly in the last fifty years in most countries. Almost universal literacy and numeracy rates.
Economic infrastructure – including transport, telecommunications, energy and water/sanitation	Tends to have reliability/maintenance problems related to expenditure allocation and cost recovery issues in particular.	Tends to be very well developed, operated and maintained with relatively little problems over financial sustainability – and widely supports 'just in time' stock control systems in production and marketing.
Research and development system	Tends to be limited to routine 'testing' and 'quality control' activities, with a relatively low average level of technical skills in the labour force and limited laboratory/experimental capacity.	Covers three categories of activity – routine, adaptive and basic research, with a relatively high average level of technical skills and significant laboratory/experimental capacity.
Economic and other statistics	Extremely limited resources for collection and verification of statistics so that the number of series available and the accuracy of those that are available make economic analysis difficult. Significant need to generate original survey data for research and policy analysis.	Well-developed and established statistical systems with wide availability of data banks as a basis for research and policy analysis.

continued

Box 2.2 (continued)

Characteristic	*'Developing countries'*	*'Developed industrial countries'*
Commodity and factor markets	Informal markets are widespread with comparatively little regulation.	Tend to be highly structured and are regulated by both government and self-generated industrial arrangements.
Legal infrastructure	Property rights widely difficult to establish; contracts widely difficult to enforce effectively.	Contestable markets with legal enforceability are almost universal.
Significance of 'subsistence' or non-monetary production	The significance of 'subsistence' production (i.e. production for self-consumption – or household production for own consumption) is still high in many developing countries – particularly in rural areas and especially in agriculture. This has important ramifications for economic behaviour (see Chapter 4).	Although non-monetary economic transactions are still significant, they account for only a small proportion of the economy and do not affect economic behaviour to any great extent.

Notes:

a) This 'stylised' comparison of some of the economic characteristics of developing countries and developed industrial countries necessarily neglects the considerable diversity within each of these categories and the fact that, in reality, the comparison would need to be more of a 'continuum'. Not the least of the differences, in addition to per capita income levels, are those relating to the physical size of countries and to the population level.

b) It can be noted that the 2008–09 global banking and financial crisis was caused largely by inadequate bank regulation and supervision in developed industrial countries, while the 1997 Asian financial crisis was caused largely by limitations of the financial sector in South East Asian countries (for further discussion see Chapters 6 and 9 of this book).

approach was also used by Dornbusch and Helmers (1988), and emerged in a revised (and evolving) form by the World Bank as their RMSM (Revised Minimum Standard Model) for macroeconomic modelling (for example, Chen et al., 2004; World Bank, 1998b). The literature on macroeconomic theory for developing countries was further enhanced by the publication of major books by Sachs and Larrain (1992) and by Agénor and Montiel (1999), and other works particularly concerned with the macroeconomics of 'structural adjustment' (including Tarp, 1993). This is indicative of the fact that professional economists, even in the international financial institutions (IFIs – the World Bank and the IMF in particular), have given a significant amount of attention to the extent to which conventional economic theory needs to be adapted to fit the characteristics of developing countries.

2.4 The significance of paradigms, schools of thought and 'disciplines'

There has been increasing awareness of the significance of clashes between different paradigms in the development of the intellectual content of disciplines over the last few decades, and nowhere is this clearer than in the literature relating to development studies (Sumner and Tribe, 2008a: particularly Chapter 3). Paradigm clashes or shifts can be associated either with major shifts in the conceptualisation of a discipline or subject area, or with significantly different approaches to the subject matter of a discipline by different schools of thought. For example, the discussion of economic growth theory in Chapter 4 of this book highlights a contrast between the approaches of traditional neo-classical economists and those embracing 'endogenous growth' – a contrast which can be characterised as a clash between paradigms or schools of thought. Where such a clash occurs, it is quite possible for communication between proponents within the different schools of thought to break down. It is not possible to refute the propositions of one school of thought on the basis of the intellectual position of the other school of thought.

At a less drastic level, there will always be evolution of thinking and of intellectual approaches within disciplines over time. Perhaps one of the clearest discussions of this within development economics is by Toye in his two editions of *Dilemmas of Development* (1987 and

1993), and in his overview chapter in Chang's recent edited book (2003). Toye distinguishes between clear phases within economic thinking about international development – based partly on intellectual evolution and partly on contrasts between the ideological positions taken by different schools of thought.

In addition to changes and conflicts within development economics (and within economics more generally), it is also necessary to be aware of the factors affecting the inter-relationship between disciplines. A collection of essays exploring the diversity of disciplinary views on development was edited by Leeson and Minogue in the late 1980s (1988). In this collection, chapters by economists, political scientists and sociologists/anthropologists combine to provide a much broader and richer analysis of international development than could be provided with a single discipline. While some writers have been concerned about whether different disciplines speak with each other (Cosgel, 2006), there also needs to be concern over whether different economic schools of thought speak with each other.

Box 2.3 has reproduced an extract from a long paper by Bardhan, a highly respected economist who has worked within the World Bank as well as within the academic world. Bardhan has a chapter which reviews the state of development economics in a major publication by Elsevier (1988), but more recently has concerned himself – inter alia – with the relationship between economics and anthropology (Bardhan and Ray, 2006). The extract in the box is very revealing in describing the economist's approach to empirical study, as well as the types of criticism to which it has been exposed by anthropologists in particular. It can be noted that Bardhan is not alone in bridging the gap between the World Bank and universities. A considerable number of academic economists have worked at senior levels within the economics division of the World Bank – including Joseph Stiglitz, Nicholas Stern, Ravi Kanbur and Paul Collier in recent years.

Finally, the journal *World Development* was notable in publishing a range of articles in 2002 reviewing a considerable number of issues which are the concern of this chapter. Kanbur (2002) introduces a small symposium of contributions (by Harriss, Jackson and White) which explore the extent to which individual disciplines contribute to the cross-disciplinarity of 'development studies'. Jackson (2002) describes the problems of cross-disciplinarity in the study of gender in

---

Box 2.3

Bardhan and Ray's views on cross-disciplinarity

'Outcomes in economic analysis have two characteristics – they serve as predictions (including predicting backward to understand changes that took place in history), and (when possible) they describe equilibrium points in the economy. Prediction is valuable in thinking about social change, and the sharp predictions of economics make it more influential in policy circles than the "softer" social sciences. But anthropologists are concerned that economists' assumptions and models are too simple to be socially useful, and that prediction of a phenomenon under a given set of constraints is too readily conflated with justification of an existing institutional set-up. Yet others argue that in situations of rapid social and economic change, only the obvious can be "predicted". … Causal explanations draw upon repeated empirical observations of the event and its supposed cause, as well as upon theories of the underlying mechanisms that supposedly produce the explained event. In economic theorizing, the causal arrow from cause C to event E is clearly specified. It is built into the model specification, and the model (in theory) stands or falls or wobbles on the basis of the accuracy of its predictions. Attributing causation in a regression analysis is a more complex matter – real data naturally create real problems. The causal arrows are not specified in statistical models, they have to be inferred from the strength and significance of the correlation between the dependent variable and the relevant independent variables. Of course, correlation on its own, however strong, cannot pass for causation.'

Source: Bardhan and Ray (2006: 6)

---

a development context. White (2002) places emphasis on the issues associated with the use of quantitative and qualitative analysis of poverty, although this highlights a major issue within the entire discussion of relative rigour and effectiveness of individual branches of the social sciences. Harriss (2002) explores the concept of the 'discipline' in the context of international development studies.

Fine (2002) responds to many of the issues raised in the articles cited in the previous paragraph. He laments the influence of a 'narrow' modern economics profession on other social sciences, writing from the 'broad' perspective of 'the political economy of capitalism' and from a development economics/development studies standpoint. A quotation provides the essence of his main argument:

> The broader implication is that crossdisciplinary endeavor that includes economics is liable to be caught on the horns of a dilemma, how to incorporate the economic without economics. If the analysis remains truly social in the sense of the systemic distinct from aggregating over individuals, then mainstream economics has very little, if not nothing to offer. For it is silent over the social relations, structures, power, conflicts and meanings that have traditionally been the preoccupation of the social sciences. This is especially important for development studies.
>
> (Fine, 2002: 2066)

It is clear that the issues addressed by this chapter have been the subject of lively debate, which is likely to be continued well into the future.

2.5 Summary

- Development economics needs to be considered within the discipline of economics and within the 'subject area' of development studies.
- For cross-disciplinarity to work within development studies it is necessary for disciplinary boundaries to be rather more relaxed and flexible than is often the case, so that the rigid application of disciplinary 'property rights' needs to be discouraged.
- The economists' view of the universality of basic economic principles and methods is broadly acceptable within development economics, but this needs to take account of the diversity of economic characteristics across the world, and particularly the differences between the economic structures of developing and developed industrial countries.
- There is a large, well-established and growing literature on development economics, with a healthy diversity of views and with evolving intellectual approaches.
- For economic analysis of developing countries, and particularly for quantitative analysis, there is a problem with the ready availability of statistical series which are up-to-date and reliable.
- A distinction can be made between 'development economics' and the 'economics of developing countries' – the former being concerned with the 'big picture' and the latter with the use of economic principles to undertake empirical economic analysis within developing countries.
- Some of the 'models' of economic development use comparatively simple economic concepts to build up very significant generalisations about the economic nature of 'development' – examples have been given from work by Lewis, Myrdal, Singer/Prebisch and Vernon.

- The controversy over whether development economics exists as a sub-division of economics, and over the role of economics within a cross-disciplinary development studies needs to be viewed within the context of a diversity of specialist areas within economics (many of them 'cross-cutting'), and within the contexts of paradigm changes and of coexistence of parallel paradigms.
- An innovatory attempt has been made to specify the nature of the differences between the structural characteristics of developing and developed industrial economies, allowing for the fact that stylised categorisations inevitably over-simplify reality.
- A view of the role of development economics within a cross-disciplinary development studies needs to allow for the contrasts, diversity of approaches and the inter-relationships within and between different disciplines – not only within the social sciences, but also including the physical sciences.

Questions for discussion

1 How well integrated is economics in the traditions and discourses of development studies, and what do you feel economics contributes to the 'world view' of international development?

2 How universal do you feel basic economic principles to be, and in particular how relevant are they to the characteristics of developing countries?

3 What are the contributions of 'models' of economic development to an understanding of long-term international development?

4 To what extent do divisions between major 'disciplines' hinder effective discourse around the issues of international development, as expressed in development studies?

5 How important do you feel interaction within and between the social sciences, physical sciences and other disciplines/subject areas to be in furthering our understanding of international development?

Suggested further reading

Fine, B. (2002) Economics Imperialism and the New Development Economics as Kuhnian Paradigm Shift? *World Development*. 30 (12) December: 2057–70.

Jomo, K.S. and Fine, B. (eds) *The New Development Economics: After the Washington Consensus.* London: Zed Books, 68–86.

Kanbur, R. (2002) Economics, Social Science and Development, *World Development.* 30 (3) March: 477–86

Sumner, A. and Tribe, M. (2008b) Development Studies and Cross-disciplinarity: Research at the Social Science-Physical Science Interface. *Journal of International Development.* 20 (6) August: 751–67.

Thirlwall, A.P. (2006) *Growth and Development with Special Reference to Developing Countries* (8th edn). Houndmills, Basingstoke: Palgrave Macmillan.

Todaro, M. and Smith, S. (2008) *Economic Development* (10th edn). London: Pearson Addison Wesley.

Toye, J.F.J. (2003) Changing Perspectives in Development Economics. In Chang, H.-J. (ed.) *Rethinking Development Economics.* London: Anthem Press, 21–40.

Tribe, M. and Sumner, A. (2006) Development Economics at a Crossroads: Introduction to a Policy Arena. *Journal of International Development.* 18 (7) October: 957–66.

Economic concepts used in this chapter

Microeconomics / macroeconomics
Neo-classical economic theory / classical political economy / heterodox economics
Economic growth and economic development / structural characteristics
Economic 'model'
Broad strategical approaches to economic development – big push, critical minimum effort, balanced vs unbalanced growth
National income / value added
Capital accumulation / capital stock / investment / profit / surplus value
Endogenous growth
External economies / inter-sectoral linkages
Modern sector / traditional sector
Unlimited supply of labour / wage rate (average and marginal) / employment
Production function / marginal product / diminishing marginal returns
Centre–periphery / cumulative inequality
Terms of trade
Competitive markets / price takers
Price elasticity of demand / price elasticity of supply / cross-elasticity (elasticity of substitution / income elasticity of demand

Foreign exchange rate / interest rate / tax rate
Product cycle
Neoliberal ideological position

Notes

1 Dudley Seers was enormously influential within the 'development' field from the mid-1950s until his untimely death in 1983. He published widely on issues of considerable contemporary interest such as poverty, employment and the interrelationship between advanced industrial and developing countries. One easy way to access a list of his publications is to search the online catalogue available from COPAC (www.copac.ac.uk).
2 It is notable that in some parts of the economic literature there is a deliberate failure to cite studies published by writers not belonging to 'accepted' schools of thought – this is analogous to 'citation clubs', but is intellectually even less defensible.

3 Economic growth and structural change

3.1 Introduction

All growing economies undergo structural change in terms of the sectoral composition of output and, to a lesser extent, in terms of the sectoral distribution of employment. It would be a strange economy indeed where all economic sectors grew at the same rate leaving the sectoral proportions within the overall economy unchanged. Although the processes of economic growth and structural change are intimately connected, this does not imply that the outcomes of these two processes are necessarily consistent with some normative concept of economic development. Structural change may well destroy existing economic activities and employment opportunities, in both less developed and developed market economies, and development itself becomes a process of creating new economic sectors and activities and developing new employment opportunities. Nevertheless, it is important to discover whether historical patterns of structural change can be identified, and whether or not lessons can be learned from such exercises.

3.2 Why do growing economies undergo structural change?

One of the most obvious long-term changes in developing economies is the relative rise of the industrial sector as a whole (which includes construction, utilities – water and sanitation, electricity and gas – and mineral extraction)[1] and of manufacturing in particular. Although the manufacturing sector is usually regarded as being fundamental to the development of a high productivity 'modern' economy, the agricultural sector plays a critically important role in economic development – as is explained in Box 3.1 in this chapter.

Structural change can be regarded as having a range of dimensions, even in a strictly economic sense. Most of the discussion in this chapter has been restricted to the characteristics of sectors and sub-sectors of the economy. However, in the long-run, other aspects of economic development relate to the scale of production, to the technological characteristics of the economy, and to the ownership characteristics (including, for example, joint-stock ownership, foreign direct investment and public/private ownership). These issues have very great significance in the context of economic development.

Not all sectors of a growing economy exhibit similar capacities for productivity growth. Although technological change in the agricultural sector has been important in the past (the Green Revolution – the introduction of dwarf, high yielding varieties of wheat and rice, along with increased use of fertilisers and irrigation), and may well be in the future if genetically modified (GM) crops become more generally accepted, it is nevertheless the case that productivity growth in agriculture will be less than that in the manufacturing sector. Nevertheless, historically the 'agricultural revolution' in the United Kingdom (including crop rotation and other science-based approaches to productivity growth) preceded the 'industrial revolution' (Deane, 1979: *passim*). Manufacturing is considered 'special' by most development economists because of its particular characteristics – rapid changes in the technological characteristics of production processes and of products, increasing returns to scale, inter-industry linkages and external economies of all kinds. Indeed, Verdoorn's law (named after the Dutch economist P.J. Verdoorn) states that there is a strong positive correlation between the growth of manufacturing output and the growth of productivity in the manufacturing sector and there is empirical evidence to support this hypothesis (Thirlwall, 2006: Chapter 3)

Rapid productivity growth in the manufacturing sector combined with changes in patterns of consumption towards manufactured goods as per capita incomes grow (Engel's Curves) together produce a growth path for the manufacturing sector's share in total output that resembles

Box 3.1

The role of agriculture in economic development

Although the agricultural sector typically declines in relative importance as per capita incomes grow, it has long been recognised by development economists that the transformation of the agricultural sector, resulting in higher rates of productivity growth and incomes, is vital. Kuznets (1971) summarised the changing relationships between agriculture and the rest of the economy in terms of three contributions to development. The *product contribution* is the increasing supply of foodstuffs to the growing non-agricultural population and the industrial crops (cotton, sisal) that are supplied to processing industries. The *factor contribution* refers to the supply of labour from agriculture to the rest of the economy and the net outflow of capital (the extraction of the surplus from agriculture), which we will refer to below. The *market contribution* refers to the market that the agricultural sector provides for domestically produced manufactured goods (furniture, processed foods, building materials, bicycles, electrical goods). Agricultural exports (coffee, tea, sugar) are also, of course, important earners of foreign exchange, which can in turn be used to purchase capital items from abroad.

More recently, the World Bank (2007d) has focused attention on the agricultural sector. It emphasises that in the twenty-first century, agriculture continues to be a fundamental instrument for sustainable development and poverty reduction. It estimates that three-quarters of poor people in developing countries live in rural areas (2.1 billion people living on less than $2 a day) and that most depend on agriculture for their livelihoods. The World Bank distinguishes between three different contexts for agriculture – agriculture-based countries, transforming countries and urbanised countries. In agriculture-based countries (most of sub-Saharan Africa) agricultural development and its associated industries are essential to economic growth and the reduction of mass poverty. In transforming countries (which include South and East Asia, the Middle East and North Africa), the World Bank recommends a comprehensive approach to shifting to high-value agriculture, decentralising non-farm activities to rural areas and helping people move out of agriculture. In urbanised countries (most of Latin America and Central Asia), the Bank argues that agricultural development can help to reduce remaining rural poverty by helping smallholders become direct suppliers to modern food markets and jobs are created in agriculture and agro-industrial activities.

For many countries, however, the basic economic problem of inter-sectoral resource allocation remains. If agriculture remains the dominant sector in the economy, the decision has to be taken as to how to extract the 'surplus' (or 'profit') from it. This can occur a) through taxing the agriculture sector, b) turning the terms of trade against the sector (lowering the prices at which farmers sell their products and raising the price of what they buy, for example), c) extracting the surplus by coercion (essentially the Soviet model) or d) by persuading farmers to put their savings into non-agricultural investments. But achieving a balance between the two sectors is difficult. If there is over-emphasis on the urban, industrial sector, farmers may react by producing less or sending less to market (or, indeed, smuggling their output to countries where the price is higher). Many neo-classical economists argue that it is mistaken government policy that is the cause of agricultural sector failure in many developing countries, especially those of sub-Saharan Africa.

an S-shaped curve (a logistic curve) as illustrated in Figure 3.1. The rate of growth of the share of manufacturing in total output will begin to slow down at higher levels of per capita income, and at much higher levels of per capita income the share of manufacturing may cease to grow or even fall. This has been referred to as 'deindustrialisation' in developed market economies such as the USA and the UK where traditional heavy industries such as coal, steel and engineering have declined in importance, or even disappeared completely, to be replaced by service sector employment (retailing, financial services, leisure activities, etc.).

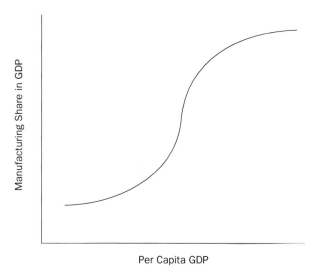

Per Capita GDP

3.1 Hypothesised growth path of manufacturing sector's share in GDP

It is also unlikely that the services sector (which contains a very wide range of economic activities) will be able to sustain high rates of productivity growth over long periods of time. If we take the example of the tourist sector, which is of key importance in many low- and middle-income economies (and which includes both international and domestic tourists), extensive growth is possible whilst there are suitable locations for new hotels, resorts, theme parks, etc., but it is difficult to see rapid productivity growth (intensive growth) occurring in this sector once extensive growth comes to an end, even though hotels and resorts can be rebuilt and go 'up-market' to attract wealthier customers. The same is true in health and education, where productivity growth may be achieved at the cost of the quality of the services provided (more students but fewer teachers; more patients but fewer nurses).

In the financial sector, technological change originating in the manufacturing sector, such as the introduction of computers, has obviously boosted productivity, but its impact on long-term productivity growth is less clear. Modern technology has applications

3.2 *Modern technology in Kenyan small-scale dairy processing*

Note: The Kitinda Dairy Farmers' Cooperative in Western Kenya employed modern ultra-heat-treatment technology in small-scale milk processing in the late 1980s. The equipment had been supplied under a Finnish aid project demonstrating the role of international aid in technology transfer.

Photograph: Michael Tribe.

within small-scale rural processing industries as well as in large-scale urban enterprises, as is illustrated by Figure 3.2. The service sector in low-income economies is so mixed (low productivity informal sector activities) that generalisations are difficult to make. In high-income economies, the service sector accounts for the major share of both national output and employment.

3.3 The work of Simon Kuznets

Simon Kuznets, an American economist, statistician and economic historian, and Nobel Laureate in 1971, was among the first economists to draw attention to the consistent patterns of structural change observed as per capita incomes increased. Mainly using data from developed countries, Kuznets grouped countries according to per capita income and demonstrated that, as per capita income rose, there was:

- a consistent decline in the share of agriculture (the primary sector) in national product;
- a consistent increase in the share of the industrial sector (the secondary sector) in national product;
- no clear pattern with respect to the share of the service sector (the tertiary sector).

Working mainly in the 1950s, Kuznets was constrained by the fact that most developing countries had very rudimentary statistical systems. It was therefore impossible to undertake systematic study of developing countries on the basis of data of a comparable standard as those available from developed countries at that time.

With respect to the share of the labour force employed in the different sectors, the following patterns emerged:

- A decline in the proportion of the labour force engaged in the agricultural sector.
- An increase, for most countries, in the proportion of the labour force in the industrial sector, but this increase was less consistent and less pronounced than its rise in the share of national output.
- A rise in the share of service sector employment in total employment, with the share of transportation and trade in particular rising most consistently.

Kuznets was prolific in publishing the results of his research in this area, but perhaps the most appropriate references are a book of collected essays (1971), an overview article in the journal *American Economic Review* (1973) and a special issue of the journal *Economic Development and Cultural Change* (1956). Subsequent empirical work tended to provide support for Kuznet's hypotheses, and many economists have argued that highlighting the major characteristics of historical patterns of development provides useful guidelines for policy makers and economic planners. Other economists argue that great care should be exercised in drawing conclusions from such work.

3.4 Industrialisation and structural change: the work of Hollis Chenery

The terms 'industry' or 'industrial sector' are used in international definitions to include mining and quarrying, electricity, gas and water and construction, as well as the manufacturing sector. Our focus in what follows is on the manufacturing sector, broadly defined to include manufacturing enterprises in the so-called 'modern' sector of the economy producing what are commonly accepted as manufactured goods.

A number of attempts have been made to estimate a 'normal' or 'standard' pattern of industrial growth. The basic hypothesis is that as per capita incomes rise, 'industrialisation occurs with a sufficient degree of uniformity across countries to produce consistent patterns of change in resource allocation, factor use and related phenomena' (Balance et al., 1982: 109).

Why should we expect to find uniformities in the transition from a 'traditional' to a 'developed' economy in all economies over any given period of time? The most important factors identified by economists include:

- similarly evolving patterns of consumer demand as per capita incomes rise, specifically a fall in the share of expenditure on basic foodstuffs and a rise in the share of expenditure on manufactured goods (including manufactured foodstuffs);
- the accumulation of capital, both physical and human, at a rate higher than the growth of the labour force, thus raising the

capital/labour ratio and increasing productivity (output per unit of labour), requiring an increase in the production of capital goods (machinery and construction);
- increased access for all countries to advanced technologies (a debateable point as noted in the discussion of globalisation);
- increased access to international trade and capital flows.

Of course, there are other factors that might be expected to produce differing patterns of structural change – differences in development policy objectives, variations in natural resource endowments, differences in country size and differences in access to foreign capital. But, for economists such as Hollis Chenery (1979), who was for a time vice president for development policy at the World Bank, in any given historical period the factors making for uniformity tend to predominate.

Chenery (1960: 635) had earlier identified what he called a 'contemporary pattern of growth'. He maintained that the close relationship between levels of income and industrial output was more pronounced than would be predicted from changes in demand alone, and that supply side factors (identified above) had to be included in any explanation of the uniformities observed in industrial growth over time. Industrialisation involved a number of changes in economic structure, including:

- a rise in the relative importance of manufacturing industry;
- a change in the composition of manufacturing production;
- changes in techniques of production and sources of supply for individual commodities.

Chenery identified the causes of industrialisation as the substitution of domestic production for imports (import substituting industrialisation or ISI), growth in final demand for manufactured consumer goods as well as growth in intermediate demand for producer goods. In other words, he argued that there was a 'normal' or 'proportional' pattern of growth, but that there were deviations from this normal pattern largely caused by changes in supply side conditions – the substitution of domestic production for imports and to a lesser extent the substitution of factory-made goods for handicraft goods and services. Other factors of significance consisted of market size (including the importance of economies of scale), income distribution and the availability of natural resources, the expectation being that countries lacking natural resources would have to turn to manufacturing at an earlier stage in their development to compensate for the lack of primary products or minerals for export and/or domestic use. Import Substitution Industrialisation (ISI) is a form of industrialisation strategy which was widely adopted by developing country governments in the 1950s and 1960s (including trade protection), but was increasingly abandoned during and after the 1970s with the onset of trade liberalisation. ISI was replaced by the Export Orientation Industrialisation (EOI) strategy. In practice private sector industrial development (including that based on foreign direct investment and joint ventures) often follows both import substitution and export orientation approaches.

In his later work Chenery (1979; Chenery and Syrquin, 1975) developed more complex models to simulate the evolution of the structure of production with rising income. A more complex interpretation of the industrialisation process, as noted above, was seen as rising from the interplay of rising consumer demands for manufactured goods, changing factor proportions, the impact of trade policies and technological development.

Table 3.1 sets out some statistics on the changing composition of national production over time (time series analysis) for groups of countries by income level. The statistics also permit a comparison of the composition of national production in particular years between groups of countries with different income levels (cross-section analysis). Data for annual growth rates have also been included in the table. The data have been taken from the World Bank's *World Development Indicators* (2007a and b), and have been included here as part of an 'evidence-based' approach.

The first thing to note about Table 3.1 is the number of blank spaces where data are not available. This even applies to high-income countries, although it is more understandable in the case of lower-income countries. Obviously, there are data for some individual countries which are available, but to refer to these would have increased the complexity of the table considerably. Readers wishing to view a wider range of data are referred to the original World Bank source (which is available from libraries, online or on CD). The World Bank (and other international organisations) take great care to check statistics received from individual reporting countries for both accuracy and consistency (of definitions etc.), but such checks are fallible.

The data in the table show very clearly how agriculture accounts for a smaller proportion of national product (i.e. total value added) over time (time series analysis) within individual countries, and for higher-income countries (i.e. international cross-sectional comparisons), because it has a lower growth rate than that for manufacturing. It can also be clearly seen that services account for a high proportion of national product at all income levels, although the composition of such services will differ considerably between lower- and higher-income countries. Services include government administration, education, health, social services, financial services, and personal household services for example.

3.5 Intra-sectoral structural change

Economists have attempted to identify patterns of structural change within the manufacturing sector itself. For example, manufacturing output includes both consumer and capital goods. Consumer goods (light industry) such as food products, textiles, leather goods and furniture are the first to develop, but metal-working industries, vehicle assembly, engineering and chemical industries (heavy industries, sometimes called capital goods) soon appear and grow more rapidly. The structure of the manufacturing sector itself thus changes in favour of heavy industry. Similar patterns of intra-sectoral structural change emerge if manufacturing activities are classified according to their income elasticities of demand and the stage of development at which they make their major contribution (Chenery and Taylor, 1968).

Three types of industries are introduced. Early industries supply the essential demands of the poorest people, use relatively simple technologies, have relatively low-income elasticities of demand (1.0 or less) and soon exhaust their potential for productivity growth (food, leather goods and textiles). Middle industries have slightly higher-income elasticities of demand (1.2–1.5) and grow rapidly as per capita income increases from relatively low levels, but do not increase their share of output beyond the middle range of per capita income (non-metallic minerals, rubber and wood products, chemicals, petroleum refining). Finally, late industries continue to grow faster than gross national product (GNP) up to the highest income levels and they typically double their share of GNP in the later stages of

Table 3.1 Sectoral composition of GDP and annual rates of growth

Agriculture		1960	1975	1990	2005
Heavily indebted poor countries (HIPC)	Value added (% of GDP)	..	37.2	35.2	31.4
	Annual growth (%)		-0.1	-0.8	4.3
Least developed countries:	Value added (% of GDP)	..	..	36.5	27.9
UN classification	Annual growth (%)		..	1.8	4.7
Low income	Value added (% of GDP)	47.1	39.7	32.4	21.5
	Annual growth (%)		5.8	3.4	5.8
Lower middle income	Value added (% of GDP)	24.6	24.3	19.2	11.5
	Annual growth (%)		3.1	4.1	3.2
Upper middle income	Value added (% of GDP)	..	13.8	10.6	6.2
	Annual growth (%)		0.9	5.9	3.0
High income	Value added (% of GDP)	..	5.4	2.8	..
	Annual growth (%)		-0.1	1.6	..
High income: OECD	Value added (% of GDP)	..	5.4	2.8	..
	Annual growth (%)		0.0	1.6	..

Industry		1960	1975	1990	2005
Heavily indebted poor countries (HIPC)	Value added (% of GDP)	..	20.2	22.6	23.9
	Annual growth (%)		2.6	-2.2	6.2
Least developed countries:	Value added (% of GDP)	..	..	20.7	26.7
UN classification	Annual growth (%)		..	-0.2	10.7
Low income	Value added (% of GDP)	17.7	22.1	26.3	28.3
	Annual growth (%)		-1.8	5.5	8.5
Lower middle income	Value added (% of GDP)	37.8	40.3	38.3	41.9
	Annual growth (%)		4.0	2.3	8.8
Upper middle income	Value added (% of GDP)	..	37.6	39.1	32.1
	Annual growth (%)		..	2.5	5.2
High income	Value added (% of GDP)	..	37.1	32.4	..
	Annual growth (%)		-3.8	3.2	..
High income: OECD	Value added (% of GDP)	..	36.6	32.1	..
	Annual growth (%)		-3.8	3.0	..

Manufacturing

Manufacturing		1960	1975	1990	2005
Heavily indebted poor countries (HIPC)	Value added (% of GDP)	..	..	12.7	10.3
	Annual growth (%)			*-1.6*	*5.0*
Least developed countries: UN classification	Value added (% of GDP)	..	..	10.7	11.5
	Annual growth (%)			*1.3*	*7.9*
Low income	Value added (% of GDP)	12.5	14.2	15.3	15.2
	Annual growth (%)		*1.2*	*4.7*	*9.2*
Lower middle income	Value added (% of GDP)	..	28.1	26.6	26.6
	Annual growth (%)		*8.5*	*4.9*	*10.5*
Upper middle income	Value added (% of GDP)	..	22.4	21.6	19.0
	Annual growth (%)		*0.4*	*3.9*	*5.0*
High income	Value added (% of GDP)	..	..	..	..
	Annual growth (%)				
High income: OECD	Value added (% of GDP)	..	..	..	..
	Annual growth (%)				

Services, etc.

Services, etc.		1960	1975	1990	2005
Heavily indebted poor countries (HIPC)	Value added (% of GDP)	..	42.6	42.6	44.7
	Annual growth (%)			*0.2*	*5.0*
Least developed countries: UN classification	Value added (% of GDP)	..	..	43.1	44.7
	Annual growth (%)			*1.8*	*6.1*
Low income	Value added (% of GDP)	35.3	38.2	41.4	50.2
	Annual growth (%)		*6.0*	*4.5*	*8.6*
Lower middle income	Value added (% of GDP)	37.7	35.5	42.5	46.6
	Annual growth (%)		*5.3*	*3.3*	*5.7*
Upper middle income	Value added (% of GDP)	..	48.7	50.3	61.7
	Annual growth (%)		*5.3*	*3.3*	*5.7*
High income	Value added (% of GDP)	..	57.5	64.8	..
	Annual growth (%)		*2.0*	*3.0*	..
High income: OECD	Value added (% of GDP)	..	58.0	65.1	..
	Annual growth (%)		*1.9*	*2.9*	..

Source: World Bank: World Development Indicators (2007a).

industrialisation (clothing, printing, paper and metal products and consumer durables with high-income elasticities of demand).

Needless to say, we have to be very cautious in making such generalisations. Change within sectors (intra-sectoral change) is often more significant than the rise or fall of those sectors themselves. Witness the rise in the relative importance of goods such as prepared meals (both cooked-chilled and frozen), ice cream, confectionery and snack products (crisps for example) within the food sector. Of equal importance are the rapidity of changes in production technology and the rapid introduction of new products (the digital revolution) which shorten product life cycles (see Chapter 6 on globalisation). This means that the importance of geographical space is reduced, and the time that it takes for information about new processes and products to travel between countries is reduced (Information and communication technology – ICT). Even poorer people in low-income economies today have access to new process and product technologies (mobile phones) that did not even exist before the last decade of the twentieth century. The concept of 'jumping technologies' is relevant here and refers to circumstance where developing countries may miss out one or more stages in technological development and move straight from a 'traditional' technology to an 'advanced' technology. One example of this is abandoning the expansion of landline telephone networks and going straight to mobile telephony as has been further discussed in Box 3.2. The role of transnational corporations (TNCs) in developing and transferring new products and processes means that many countries, for example Malaysia, assemble and export high-technology products, changing our patterns of growth hypotheses. Globalisation involves the relocation of certain manufacturing processes to low- and middle-income economies (automobile components, electronics, semiconductors) and has produced very different patterns of growth that have not yet been fully explored. For example, advanced technology, high-income products (mobile phones, laptop and desktop computers) are assembled in and exported by these countries because of the transnational production networks of giant manufacturing companies. But the technology is not 'produced' in these countries and the production and export of these commodities is not consistent with their per capita incomes.

Table 3.2 includes some statistics on the composition of manufacturing production (value added) for a range of low-, middle- and high-income countries, again taken from the World Bank's *World*

Box 3.2

New technologies: the use of mobile phones in Africa

The use of mobile phones is transforming commerce, health care and social lives in Africa, according to the United Nations (UNCTAD, 2009b). Mobile subscriptions in Africa rose from 54 million in 2003 to almost 350 million in 2008, the fastest growth in the world. Gabon, the Seychelles and South Africa had almost 100 subscriptions per 100 inhabitants. In North Africa, the average penetration stood at almost two-thirds of the population, and for Africa as a whole, it was over one-third. In Uganda, for example, mobile phone penetration rose between 1995 and 2008 from 0.2 to 23 per 100 inhabitants, and there have been significant investments in infrastructure, especially in rural areas. More widespread use of mobile phones has led to the emergence of new services and applications. For many small and medium-sized enterprises (SMEs), the mobile phone has become more important than the computer as the most important ICT tool. African countries are pioneering mobile banking and electronic transactions services. Many farmers use their mobiles to trade and check market prices.

In contrast to the use of mobile phones, African economies are lagging behind other developing regions in internet use and broadband connectivity. Broadband use in Africa is highly concentrated, with five countries (Algeria, Egypt, Morocco, South Africa and Tunisia) accounting for 90 per cent of all subscriptions.

The above examples highlight two points: rapid technological change leads to equally rapid changes in patterns of consumption, but rapid technological change leaves many countries behind and that gap has implications for the economic growth and competitiveness of those countries.

Development Indicators (2007a). Data are not available at this level of detail for groups of countries. Chemical production is an important part of a modern economy, and some of the low-income economies hardly have a chemicals sub-sector at all. However, it is notable that India (as a large low-income economy) does have a very significant and growing chemicals industry. Another sub-sector which is an essential part of a high-income industrialised economy is the machinery and transport equipment sector, and India is again untypical of low-income economies in having a large amount of its manufacturing output in this area. It can easily be seen that low-income economies hardly feature in this sub-sector at all. It is also clear that the food, beverages and tobacco sub-sector and the textiles and clothing sub-sector contribute a much higher proportion of manufacturing output at lower levels of income, as would be expected (and predicted) based on the earlier discussion in this chapter.

Table 3.2 also shows the proportion of manufactured exports in total merchandise exports. Merchandise exports include agricultural products and other raw material and mineral products. Two particular trends are very noticeable for this last section of the table. First, the impact of the export orientation of developing countries is shown – India has 76.5 per cent of her exports in the manufacturing category in 2000, Indonesia 57.1 per cent, Malaysia 80.4 per cent and Brazil 58.5 per cent. Second, the effect of regional trading/economic arrangements is very clear – both for Europe and for North America. By 2000 Germany, Spain and Sweden had around 80 per cent of their exports in the manufactures category, while Mexico had reached the same level by that year. The transformation of patterns of international trade in manufactured products is shown very clearly by the statistics.

3.6 Patterns of industrialisation: an evaluation

Opinions have always differed as to the value and relevance of statistical analyses that attempt to identify 'normal' patterns of economic growth and industrialisation. On the one hand, it is argued that such analyses more clearly highlight the salient features of historical patterns of development and provide useful guidelines for policy makers (Colman and Nixson, 1994).[2] Others argue that the statistical evidence is ambiguous and that care should be exercised in drawing conclusions from these analyses. In the more recent period of neo-liberalism, orthodox economists argue that it is the market that should decide on what should be produced and exported, and the factor proportions used in production, rather than the government or policy makers.

Clearly the pattern of development and industrialisation of each individual country will be influenced by its own economic and political history, its relationships with other countries (especially its trade with the developed market economies), the role that the state plays in the development process, at least in its early stages, and the global economic and political environment. With these important qualifications in mind, the identification of 'normal' patterns of

Table 3.2 Percentage of manufacturing value added originating in specific sub-sectors and manufactured exports as a percentage of total merchandise exports (for selected countries)

	Chemicals				Food, beverages and tobacco				Textiles and clothing				Machinery and transport equipment				Manufactured exports as % of merchandise exports			
	1970	1980	1990	2000	1970	1980	1990	2000	1970	1980	1990	2000	1970	1980	1990	2000	1970	1980	1990	2000
Ethiopia	–	–	0.2	0.4	–	–	12.1	4.7	–	–	5.4	2.7	–	–	0.1	–	–	–	–	9.8
India	13.6	14.1	13.9	21.1	12.8	9.1	11.7	13.2	21.0	21.3	15.2	12.6	20.2	25.1	25.5	19.1	51.9	58.6	70.4	76.5
Ghana	4.4	4.5	–	7.3	33.9	37.1	–	31.8	16.2	10.8	–	4.9	4.4	1.9	–	0.6	0.7	1.0	–	14.7
Indonesia	6.0	11.4	8.9	10.5	65.5	31.8	27.5	18.0	13.8	13.3	14.6	16.7	1.8	13.3	11.8	24.1	1.2	2.3	35.5	57.1
Malaysia	8.6	5.4	10.8	7.9	25.8	23.8	13.2	8.0	3.5	7.0	6.5	4.1	8.1	20.3	30.7	41.5	6.5	18.8	53.8	80.4
Mexico	–	–	17.6	15.4	–	–	21.6	25.4	–	–	4.8	3.9	–	–	24.2	27.1	32.5	11.9	43.5	83.5
Brazil	–	–	–	–	–	–	13.6	–	–	–	12.2	–	–	–	26.6	–	13.2	37.2	51.9	58.5
Philippines	12.7	14.1	11.8	11.2	38.8	30.3	38.9	37.6	8.3	12.8	10.7	9.4	8.0	12.2	12.6	8.6	7.5	21.1	37.8	91.7
Spain	11.1	8.7	10.3	9.6	12.9	15.9	17.9	14.2	15.1	11.5	7.9	6.7	16.2	23.1	25.1	22.9	53.4	71.7	74.9	77.7
Sweden	5.5	7.2	8.8	11.2	10.1	10.2	10.2	7.3	5.7	3.0	1.7	1.0	29.6	32.9	32.5	38.9	72.1	77.8	82.6	85.1
United States	10.4	9.7	11.7	–	11.6	10.6	12.4	–	8.0	6.2	4.9	–	30.8	33.6	31.1	–	66.5	65.7	74.7	83.9
Germany	–	–	–	–	–	–	–	–	–	–	–	2.3	–	–	–	40.7	86.8	84.3	89.1	84.0

Source: World Bank: World Development Indicators (2007a).

Note: – = not available.

structural change and industrialisation can be of use for policy makers, not in the sense that they provide rigid guidelines as to what 'should' be done, but rather in the sense that they raise important issues relating to the nature, consequences and management of structural change and permit a more informed discussion of these issues. Box 3.3 outlines some contributions to discussion about the stages of economic diversification.

3.7 Structural change and international trade

As the structure of domestic production undergoes the sort of changes that we have described above, we would expect to see similar, although not identical, changes in the structure of the individual economy's international trade, although with variations across countries depending on their size and natural resource endowments.

As the economy begins to diversify its productive base through the establishment of manufacturing enterprises, for example, the composition of imports is likely to change. Under import substituting industrialisation (ISI) regimes, domestically produced consumer goods are substituted for their imported equivalents, with the latter often being prohibited or subject to high import tariffs and duties. The share of consumer goods imports (shoes for example) in total imports may thus fall, to be replaced by imports of machines (investment goods) to make those shoes, and perhaps the raw materials or intermediate goods (plastics, leather and metal buckles, for example) that go into their making. At a later stage in the process of industrialisation, it is quite possible that countries will begin to import capital goods (including machines that make other machines) in order to establish their own machine-making industries. For such import substituting economies, the composition of imports changes but trade does not become less important to the economy. Many domestically produced goods remain highly import intensive.

Another example is the country that pursues the export of manufactured goods (export orientation industrialisation – EOI). In this case, the composition of both imports and exports is likely to change. By definition, manufactured goods exports will become a larger proportion of total exports, but again, those exports are likely to be very import intensive (firms in the domestic economy will not

Box 3.3

Stages of diversification

Orthodox trade theory (see Chapter 7) suggests that as countries open up to trade they specialise in the production of a specific range of goods according to their comparative advantage. At low per capita incomes, countries tend to specialise in the production and export of commodities based on their natural resource endowments – for example, coffee, cotton, bananas or minerals. But recent empirical evidence suggests that as countries grow, sectoral production and employment, both between and within sectors, become less concentrated and more diversified and that this diversification continues until relatively late in the development process, that is, at a relatively high level of per capita income (Imbs and Wacziarg, 2003).

Imbs and Wacziarg (2003) use data from the International Labour Office (ILO) on employment shares across sectors, from the United Nations Industrial Development Organisation (UNIDO) on value added in the manufacturing sector, and from the Organisation for Economic Co-operation and Development (OECD) on both employment and value added for a number of developed market economies. According to all the measures of concentration that Imbs and Wacziarg (2003) calculate, they find that sectoral diversification goes through two stages – first one of increasing diversification and then one of increasing concentration – tracing out a U-shaped relationship when income per capita is plotted against a measure of sectoral concentration. The implication of this, of course, is that both poor and rich countries tend to have fairly concentrated patterns of production and employment, although the richer countries are less concentrated than the poor, and that it is the lower-income, rapidly growing economies that will undergo equally rapid diversification. Countries diversify over most of their development path, although smaller open economies (for example, Singapore) tend to specialise, that is, be less diversified, at lower levels of per capita income.

This pattern of development is contrary to the predictions of orthodox trade theory. Rather than countries concentrating on what they do best, economic development requires that countries develop the capability to do 'lots of other things well', that is, they develop new competitive advantages. As the eminent Harvard economist Dani Rodrik (2007) points out, the driving force of economic development cannot be the driving forces of comparative advantage as usually understood. Some countries (for example, the East Asian 'Tigers') have succeeded spectacularly in pursuing economic growth and diversification. Other countries, such as the majority of sub-Saharan economies, have yet to learn the lesson. Many economists, including Rodrik, point to the importance of sound industrial policy in achieving this transition.

produce the inputs needed for the exports, or will not achieve the quality required), leading to changes in import composition. The gross value of the exports of manufactured goods by many low-income economies obscures the relatively limited value added that they capture domestically and exaggerates the actual foreign exchange earnings (i.e. net earnings). This is especially the case where countries have established export processing zones (EPZs) – for example, Mexico, Mauritius, China – in order to attract foreign direct investment (FDI) that will bring in labour-intensive, low-value-added and import-intensive assembly and processing activities, such as the sewing of garments or the processing of data. Sanjaya Lall, in particular, emphasised the fact that technology is embedded in the characteristics of traded products, and that some countries rely principally on the export of 'traditional' types of products (e.g. basic textiles and clothing) while others focus on exports of 'advanced' products (e.g. high technology electronics) (Lall et al., 2006).

Economic growth and structural change thus have very important implications for a country's trading position and performance. It must encourage competitiveness through innovation and productivity growth in order both to maintain its position in global markets and to ensure that those enterprises serving the domestic market can remain competitive without long-lasting protection. But the structures of both exporting and import competing sectors will change over time as new competitors enter the market, and tastes and technology also change, especially with the introduction of new products (mobile phones, laptop computers).

3.8 Summary

- All growing economies undergo structural change.
- Manufacturing is the sector most likely to maintain rapid productivity growth over long periods of time.
- There is clear empirical evidence of consistent patterns of inter-sectoral structural change over time.
- Attempts have been made to identify a 'normal' pattern of industrialisation and explain deviations from it.
- Intra-sectoral change within the manufacturing sector is very important.
- Rapid technological change and the activities of TNCs are leading to new patterns of global production and consumption.

● The identification of 'normal' patterns of growth and industrialisation may help policy makers deal with the problems raised by structural change.

Questions for discussion

1 Why does the structure of an economy change as per capita income rises?

2 What can today's less developed countries (LDCs) learn from the historical experience of the now developed market economies?

3 What role does the agricultural sector play in the development process? Should agricultural sector development be given higher priority by governments in LDCs?

4 What changes might we expect to occur within the manufacturing sector as per capita incomes rise?

5 Why is the manufacturing sector considered to be 'special'?

Suggested further reading

Chenery, H.B. 1960. Patterns of Industrial Growth. *American Economic Review*. 50 (4) September: 624–54.

Chenery, H.B. 1979. *Structural Change and Development Policy*. Oxford: Oxford University Press for the World Bank.

Chenery, H.B., Robinson, S. and Syrquin, M. 1986. *Industrialisation and Growth: A Comparative Study*. Oxford: Oxford University Press for the World Bank.

Colman, D. and Nixson, F. 1994. *Economics of Change in Less Developed Countries* (3rd edn). London: Harvester Wheatsheaf.

Thirlwall, A.P. 2006. *Growth and Development with Special Reference to Developing Countries* (8th edn). Basingstoke: Palgrave Macmillan.

Useful websites

www.unido.org

Economic concepts used in this chapter

Economic growth – specialisation, diversification
Economic sector – industry, manufacturing, agriculture, services
Productivity
Green revolution
Technological change
External economies – inter-industry linkages
Increasing returns to scale
Verdoorn's law
Engel's curves
Logistic curve
De-industrialisation
'Normal' patterns of growth
Import substituting industrialisation
Export-oriented industrialisation
Export processing zone
Direct foreign investment
Light industry – heavy industry
Globalisation

Notes

1 The determination of 'sectors' of the economy is laid down in the United Nations System of National Accounts (UN SNA) (UN, 1993; World Bank, 2007a: 376). For the classification of industrial sectors in the national economy the International Standard Industrial Classification (ISIC) is used and for the classification of products in international trade the Standard International Trade Classification (SITC) is used.

2 In Colman and Nixson (1994) the relevant discussion of this issue is in Chapter 9 (including further references).

4 Economic growth and developing countries

4.1 Introduction

Within twenty years of the end of the Second World War two noted authors, from distinctly different ideological perspectives, emphasised the significance of savings and capital investment for economic growth (Lewis, 1954: 155 and 1955: Chapter V; Rostow, 1960a: 37 and 1960b: Chapter XII). Arthur Lewis, a Nobel Laureate in Economics in 1979, was born in the West Indies, but spent much of his academic career in the United Kingdom at the University of Manchester. He was the author of several very significant books and articles which had a profound influence on economic development policy just before, and for some years after, the independence of many former British colonies, particularly in sub-Saharan Africa. Walt Rostow, a noted economic historian, introduced the concept of 'stages of economic growth' (including the take-off into sustained economic growth) which reverberated around the development studies community, and provided a readily understandable strategic approach for politicians. One of the objectives of this chapter will be to explain in more detail why economists consider savings and capital investment to be so important for economic growth.

A quotation from Lewis will illustrate the significance of this issue:

> The central problem in the theory of economic development is to understand the process by which a community that was previously

> saving and investing 4 or 5 per cent of its national income or less,
> converts itself into an economy where voluntary saving is running
> at about 12 to 15 per cent of national income or more. This is the
> central problem because the central fact of economic development
> is rapid capital accumulation (including knowledge and skills with
> capital). We cannot explain any 'industrial' revolution (as the
> economic historians pretend to do) until we can explain why
> saving increased relatively to national income.
>
> (Lewis, 1954: 155)

Many serious students of international development studies might ask why economic growth is considered to be so important. At a time in global development when many would question the justification for an emphasis on economic growth on environmental grounds, it should be clear that if the Millennium Development Goals (MDGs – United Nations, 2009a) are to be achieved to an acceptable degree, then economic growth is a necessary, but not a sufficient, condition. In other words, the achievement of higher standards of living across society as a whole requires economic growth. This implies that development studies specialists concerned with the welfare objectives embodied in the MDGs need to be aware of the economic factors associated with economic growth.

A controversial but influential contribution to the discourse relating to poverty reduction is that of Dollar and Kraay (2002; 2004). However, their 'Growth is Good for the Poor' suffered from a number of shortcomings, notably the absence of a rigorous definition of poverty ('the poor') and of a clear methodological distinction between cross-section and time-series analysis (Amann et al., 2006). Despite the heated controversy generated by the Dollar and Kraay paper, the logical connection between economic growth and poverty reduction – both absolute and relative – would not be disputed by most economists.

The definition of poverty, and of 'the poor', is explored systematically in Chapter 9 of this book. The significance of the distinction between 'cross-section' and 'time-series' analysis requires a little more clarification. In principle, the problem is that in predicting the future course of events in one particular country (time-series analysis), it is not always acceptable to base predictions on observation of patterns experience across a range of other countries at one particular point in time (cross-section – or cross-country – analysis). For example, if a particular country has an average per capita income of US$500 and 50 per cent of its population is below the US$1.00 per day poverty line, and cross-country analysis shows that a sample of countries with an average per capita income of US$1,000 have an average of 30 per cent of their population below the US$1.00 per day poverty line, it does not necessarily follow that when the particular country in question reaches a per capita income level of US$1,000 that it will also have 30 per cent of its population below the US$1.00 per day poverty line.

One problem associated with the discussion of economic growth is the identification of the direction of causation between key variables. For example, are higher expenditure on education and a higher proportion of population having educational qualifications causes or consequences of economic growth? Are the higher levels of expenditure on education possible only when the national income, personal disposable incomes, and government expenditure have themselves attained higher levels – or is it higher educational attainment which causes or enables growth in the key economic variables?

A similar logical problem applies to the connection between higher levels of expenditure on medical services and improved health conditions and economic growth. Is it higher levels of income which make it possible to commit more resources to improving health, or do improved health conditions contribute towards the attainment of higher levels of productivity which in turn lead to higher levels of income?

In reality we are dealing with circular causation. Improvements in educational attainment and in health conditions are both enabled by and lead to economic growth through a type of 'virtuous circle'. Close statistical association between the variables with which we are concerned do not demonstrate 'causation'. The causation is provided by logical, rather than by statistical, analysis.

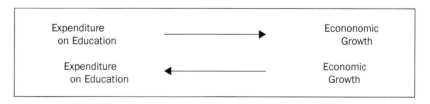

4.1 Education and economic growth – what is the direction of causation?

Another significant issue concerns the time period over which economic growth is observed and analysed. For economists, the phenomenon of 'growth' is one which relates to the 'long-run'. Contemporary economists might have forgotten, or never have been aware, that John Maynard Keynes famously referred to the fact that 'in the long run we are all dead'. Although very relevant to short-run perceptions within a few years, this adage is less relevant to growth economics, and particularly development economics, which are concerned with issues crossing decades, generations and centuries rather than with the short run.

In the short run, over periods of up to perhaps ten years, there are often fluctuations in economic activity with variations in the degree of capacity utilisation. In the longer run, changes in capacity utilisation become less significant because an 'average degree of capacity utilisation' is more important. Economic growth is concerned with the underlying long-run growth of economic capacity rather than with short-run variations in its utilisation. Harrod and Domar, who will be referred to in the discussion which follows, focused their contributions to economic thought on both the economic forces leading to the generation of cycles of activity (variations in capacity utilisation) and the growth of underlying economic capacity. More recent economic analysis in developed countries has led to a greater ability for government economic management to avoid serious fluctuations in economic activity.

The second section of this chapter focuses on theories of economic growth based on capital accumulation. The third explores the roles of technology, labour force enskillment and institutional change in 'new' theories of economic growth. The fourth section discusses the distinction between partial and total factor productivity growth, while the final section outlines some of the main conclusions from the contents of the chapter.

4.2 Theories of economic growth based on capital accumulation

In the 1930s and 1940s there was considerable concern amongst economists with the avoidance of the types of economic cycles which had led to the Great Depression and to the rise of fascism in Europe. The response of economic theorists took two principal forms. The first was focused on short- and medium-period macroeconomic management for which the main contribution was provided by John Maynard Keynes (1936). The second was concerned with the avoidance of long-term economic cycles (periodic booms and slumps) for which the main contributions were provided by Harrod (1939) and Domar (1946). For the theory of economic growth, the 'Harrod–Domar' approach (which is one part of the theory of economic cycles) provides a persuasive and simple basis for the understanding of the main factors leading to economic growth. The application of the Harrod–Domar approach to the analysis of economic growth, particularly in developing countries, is controversial and was certainly not intended by the original authors. The sections which follow use the approach as an heuristic device, and a number of *caveats* are highlighted. Two outstanding discussions of the analysis of economic growth in developing countries are by Kenny and Williams (2001) and by Fine (2006a).

The basic Harrod–Domar 'capital coefficient' model

It is appropriate to explain the fundamental features of the Harrod–Domar approach before attempting further elaboration. We start with some basic macroeconomic notation which relates to one year at a time.

$$Y \equiv C + I \quad \text{OR} \quad Y_{EXP} \equiv C + I \tag{4.1}$$

$$Y \equiv C + S \quad \text{OR} \quad Y_{DISP} \equiv C + S \tag{4.2}$$

Where:

C   = Aggregate Consumption
I   = Investment
Y   = National Income
Y_{EXP} = National Expenditure
Y_{DISP} = National Disposable Income

Equation (4.1) simply states that national income is equivalent to the sum of consumption and investment expenditures. Equation (4.2) simply states that national income is committed either to consumption or to savings.

Then $\Delta Y_{EXP} \equiv \Delta Y_{DISP}$ and $I \equiv S$ \hfill (4.3)

Equation (4.3) states that in national income accounting terms there is a necessary ex post identity between expenditure and disposable income, and between investment and savings.

$$I = \Delta K \hfill (4.4)$$

Where:

K $\quad=\quad$ Total Capital Stock
$\Delta K \quad=\quad$ Change in Total Capital Stock

Equation (4.4) states that investment in any one year is equal to the increase in the size of the capital stock from one year to the next.

It follows that if $I \equiv S$ and $I = \Delta K$ then $S \equiv \Delta K$. In other words, if investment is identical to savings, and if investment is equal to the increase in the size of the capital stock, then savings are equivalent to the increase in the size of the capital stock.

It then follows that:

$$\Delta Y = 1/\alpha \, . \, \Delta K \quad \text{or} \quad \Delta Y/\Delta K = 1/\alpha \hfill (4.5)$$

Where:

$\Delta Y \quad=\quad$ Change in Level of National Income

$\alpha \quad=\quad$ Incremental Capital/Output Ratio (ICOR)

Equation (4.5) simply states that the change in the level of national income from one year to the next is dependent upon the change in the size of the capital stock (i.e. on the level of savings and investment) and on the fundamental relationship between changes in the size of the capital stock and in the level of income (or 'stream' of income) produced. It should be noted that the incremental capital output ratio (α) is defined as $\Delta K/\Delta Y$.

If the incremental capital–output ratio (ICOR or $\Delta K/\Delta Y$) is equal to the average capital–output ratio (ACOR or K/Y), then:

$$K/Y = \Delta K/\Delta Y \hfill (4.6)$$

In practice, it is very difficult to measure the incremental capital–output ratio, and so it is generally assumed that the incremental ratio is the same as the average ratio. Having set out the basic features of this approach to the analysis of economic growth, it is possible to move on to a numerical example which will allow us to draw some important conclusions.

If the incremental capital–output ratio (ICOR) is 3:1 (meaning that three units of value in the capital stock are needed to produce one unit of income in any one year – a reasonable assumption), then:

$$Y = 1/3 \cdot K \text{ and } \Delta Y = 1/3 \cdot \Delta K \text{ and if } S = I = \Delta K$$

then

$$\Delta Y = 1/3 \cdot I \tag{4.7}$$

Equation (4.7) states that if the average relationship between changes in the size of the capital stock is the same as the incremental relationship at 3 to 1, and if savings, investment and the change in the size of the capital stock are identical, then the change in the level of income will be equal to one-third of the level of investment (which is identical to the level of savings and to the change in the size of the capital stock, Box 4.1).

For some developing countries there are issues over the comparability of investment and national income statistics where there is a significant non-monetary (subsistence) element. For the agricultural sector – which accounts for a high proportion of national output in many developing countries – the capital–output ratio is usually low anyway, so that non-monetary investment leading to marketed produce would tend to lower the capital–output ratio still further. The capital–output ratio for the infrastructure sector tends to be high, so that countries with high levels of investment in that sector would have a higher capital–output ratio.

If we divide both sides in equation (4.7) by Y, we then arrive at equation (4.8):

$$\Delta Y/Y = 1/3 \cdot I/Y \text{ and if } I \equiv S \tag{4.8}$$

then

$$\Delta Y/Y = 1/3 \cdot S/Y \text{ or } \Delta Y/Y = 1/3 \cdot \Delta K/Y \tag{4.9}$$

Box 4.1

Empirical estimates of the capital–output ratio

There has been considerable controversy over the estimation of the overall average capital–output ratio, and one of the most recent estimates has been reproduced below. It can be seen that the lowest value is 1.8:1 and the highest is 9.8:1. However, sixteen of the thirty values in the table lie between 2.5:1 and 4.0:1 – giving some credence to the use of the value of 3.0:1 for the discussion in this chapter.

Country	1970–79	1980–89	1990–97
United States	3.4	3.6	2.8
Japan	4.9	4.6	6.5
India	3.7	2.8	2.9
Indonesia	2.2	3.2	3.0
Argentina	4.8	9.8	2.0
Brazil	2.2	4.2	3.4
Venezuela	5.6	6.7	2.8
South Africa	2.7	2.8	3.7
Ivory Coast	2.6	4.3	1.8
Kenya	2.3	3.7	4.0

Source: Perkins et al. (2001: 49, Table 2.5) – calculated from the World Bank's World Development Indicators (1999).

For some developing countries there are issues over the comparability of investment and national income statistics where there is a significant non-monetary (subsistence) element. For the agricultural sector – which accounts for a high proportion of national output in many developing countries – the capital–output ratio is usually low anyway, so that non-monetary investment leading to marketed produce would tend to lower the capital–output ratio still further. The capital–output ratio for the infrastructure sector tends to be high, so that countries with high levels of investment in that sector would have a higher capital–output ratio.

Equations (4.8) and (4.9) state that the rate of growth of national income will be equal to one-third of the ratio of investment to national income or to one-third of the ratio of savings to national income.

If the rate of saving and investment in the economy (i.e. S/Y or $\Delta K/Y$) is 9 per cent, then the rate of growth of national income would be 3 per cent per annum:

$$\Delta Y/Y = 1/3 \cdot S/Y \quad \text{and} \quad \Delta Y/Y = 1/3 \cdot 9\% \tag{4.10}$$

with the result that:

$$\Delta Y/Y = 3\% \tag{4.11}$$

The main implication of this part of the explanation of the Harrod–Domar (or capital coefficient) approach to economic growth is that if the rate of saving is 9 per cent of national income, then the rate of growth of national income will be 3 per cent per annum.

This is, of course, a rather simplistic and mechanistic approach to the complex phenomenon of economic growth. However, one of the main roles of theory is to impose simplicity in order to permit a clearer understanding of complex phenomena – the process of abstraction implies that the theory will not usually be a complete representation of reality. One of the most recent criticisms of the use of the Harrod–Domar model as a basis for the understanding of economic growth has been by Easterly (1999a), who was particularly concerned with its use for the estimation of foreign aid requirements; however, his arguments are rather muddled and they do not detract from the use of the model as a basis for the systematic discussion of economic growth in this chapter. Valuable discussion of the use of the capital coefficient in the analysis of economic growth can be found in Reddaway (1962: Appendix C) and Myrdal (1968: Vol. III, Appendix 3.II) for example.

Further elaboration based on the Harrod–Domar model

The simplification which has been articulated in the previous sub-section omitted one important consideration. Each year part of the Capital Stock requires replacement – this replacement investment is known as 'depreciation', and amounts to perhaps 6 or 7 per cent of national income (GDP). The World Bank's *World Development Indicators* include a series for 'Adjusted Savings: Consumption of Fixed Capital (as a percentage of Gross National Income)' (World Bank, 2007a). The ratios for 2005 are given as 12.6 per cent for the world, 8.9 per cent for the least developed countries, 10.7 per cent for sub-Saharan Africa, and 10.3 per cent for East Asia and the Pacific. On the basis of this evidence, the use of a ratio of 6 per cent to 7 per cent errs on to the conservative side. It should be noted that there is an important distinction between the economic concept of depreciation and the equivalent financial concept. In economic terms, depreciation refers to the amount of investment required to replace those parts of the capital stock which have been 'used-up' in any one national income accounting period. The financial concept is bound up in tax allowance rules and accounting conventions which are the basis of financial depreciation allowances, so that it has little economic significance.

It is necessary to distinguish between a number of separate economic concepts related to depreciation or replacement investment. Total investment in the economy in any single period of time includes three distinct elements: a) new, or net, investment which adds to the existing capital stock; b) replacement investment or depreciation which keeps the existing capital stock intact; and c) increases in working capital which consists of stocks of materials, of work in progress and of finished products. The first two of these, excluding working capital, together comprise fixed capital formation (or fixed investment). New investment, increasing the size of the capital stock, is referred to as net fixed capital formation (NFCF), and when added to replacement investment the total is referred to as gross fixed capital formation (GFCF).

If replacement investment amounts to 6 per cent of gross domestic product (GDP – the most commonly used form of 'national income') and if the incremental capital–output ratio (ICOR) is 3:1, then it will be necessary to save and invest 6 per cent of GDP each year simply to sustain a constant level of GDP. At any rate of saving and investment below 6 per cent, GDP will fall from year to year, which suggests that we are concerned with a concept of sustainable growth of national income. In order to sustain a rate of economic growth of 3 per cent per annum – as outlined above in the previous sub-section – it will be necessary for savings and investment to be 15 per cent of GDP in every year. This amounts to 6 per cent of GDP saved and invested in order to maintain the size of the capital stock, and 9 per cent of GDP saved and invested in order to increase the size of the capital stock so that national income grows by 3 per cent per annum.

i.e. Gross Investment/GDP = Depreciation + Net Fixed Capital Formation

$$= 6\% + (3/1 \cdot \Delta Y/Y)$$

$$= 6\% + (3 . 3\%)$$
$$= 15\%$$

Let us suppose that population growth is running at 0.5 per cent per annum, not untypical for a higher-income 'developed' country. In order to sustain a constant level of per capita national income it will be necessary for national income to grow at 0.5 per cent per annum and, other things being equal, this would require a further 1.5 per cent of GDP to be saved and invested.

National income growth at 0.5 per cent in order to maintain a constant level of per capita income requires investment of 1.5 per cent of GDP (i.e. 1/3 . 1.5 per cent) and with required replacement investment at 6 per cent of GDP this means that it would be necessary to save and invest 7.5 per cent of GDP simply to maintain the same level of per capita income if population growth is 0.5 per cent per annum (Table 4.1).

If population growth were to be running at 2 per cent per annum, a not unrealistic scenario for many developing countries, this would require savings and investment of 6 per cent of GDP simply in order to compensate for the population growth and to maintain a constant level of per capita income. Added to the required replacement investment at 6 per cent for the maintenance of the capital stock this means that it would be necessary to save and invest a total of 12 per cent of GDP in order to sustain a constant level of per capita national income in many developing countries.

Table 4.1 *Summary of required investment rates for sustained per capita income growth (expressed as a percentage of GDP)*

	'Typical' developed country	'Typical' developing country
Annual replacement investment	6	6
Investment required for constant per capita income	1.5	6
Investment required for 3% sustained per capita income growth	9	9
Total required annual investment	16.5	21

Source: Refer to text.

Suppose that it is desirable that the level of per capita income should be increased in order to achieve a higher standard of living and poverty reduction. If this desired rate of growth of per capita income is 3 per cent per annum, this would be associated with a required rate of savings and investment of 9 per cent per annum (as was seen in the previous sub-section of this chapter). If this is added to the 6 per cent of GDP required as savings and investment to cover replacement investment, and to the additional 6 per cent of GDP which needs to be saved and invested to compensate for a 2 per cent per annum rate of population growth, then a total of 21 per cent of GDP would have to be saved and invested in every year.

The implication of this is that developing countries with comparatively high rates of population growth need to save and invest a significantly higher proportion of GDP simply in order to achieve the same rate of growth of per capita income as higher-income developed countries, making 'catching up' much more difficult. Higher rates of saving and investment imply lower rates of consumption, so that the achievement of a more rapid rate of growth in standards of living for the future requires lower rates of consumption (the basis for the standard of living) in the present (Tables 4.2 and 4.3).

When 'savings' were mentioned in the earlier parts of this chapter, the reference was to 'national' savings (a macroeconomic concept) and not to 'individual household' savings (which is more of a microeconomic concept). Individuals 'save' for a range of purposes. Some savings are incurred on a long-term contractual basis (such as pensions contributions), but other savings are short term and relate to temporarily delayed consumption in order to 'put funds away' for later consumption (such as for holidays, weddings or lumpy expenditure such as on private cars). Only the former – the long-term savings – can realistically contribute towards the funding of long-term productive investment in the economy. However, even if such long-term savings are deposited into a bank (even in a 'savings' deposit), they may not contribute to national savings because the bank may subsequently lend the deposits to lenders who spend their borrowed money on consumption expenditure (such as holidays, private cars or other consumption goods). This discussion does not imply any value judgement about the actions of banks making such loans for consumption purposes, but is simply intended as a reminder about the need to analyse the nature of the 'savings market' carefully in terms of its economic impact.

Table 4.2 *Investment rates and economic growth rates for a sample of countries*

Country	1970–79		1980–89		1990–99	
	GFCF/ GDP	Average GDP growth	GFCF/ GDP	Average GDP growth	GFCF/ GDP	Average GDP growth
United States	25.8	2.9	20.8	−0.7	13.2	4.5
Japan	21.7	8.5	15.8	3.0	19.5	1.8
India	23.0	7.6	20.2	−0.2	12.0	2.6
Indonesia	15.8	2.9	24.4	5.9	22.3	5.7
Argentina	2.2	7.8	29.5	6.4	26.7	4.8
Brazil	33.5	5.3	15.9	3.7	29.0	1.7
Venezuela	21.4	7.2	23.1	4.2	16.0	2.1
South Africa	26.4	3.4	19.3	2.3	16.3	1.4
Ivory Coast	19.2	3.3	20.8	3.1	17.7	3.1
Kenya	28.0	4.0	20.8	−0.2	17.7	2.4

Source: World Bank (2005) World Development Indicators (averages calculated from annual data in the original source).

Table 4.3 *Per capita income growth and population growth rates for a sample of countries*

Country	1970–79		1980–89		1990–99	
	Average population growth	Average per capita GDP growth	Average population growth	Average per capita GDP growth	Average population growth	Average per capita GDP growth
United States	1.1	2.3	0.9	2.1	1.2	1.9
Japan	1.2	4.1	0.6	3.1	0.3	1.4
India	2.3	0.6	2.1	3.7	1.8	3.8
Indonesia	2.4	5.3	1.9	4.4	1.5	3.3
Argentina	1.6	1.3	1.4	−2.1	1.1	3.4
Brazil	2.4	5.9	2.0	0.9	1.4	0.4
Venezuela	3.4	0.5	2.8	−2.9	2.1	0.3
South Africa	2.2	1.1	2.5	−0.2	2.2	−0.8
Ivory Coast	4.0	3.4	3.7	−3.8	3.1	−0.5
Kenya	3.6	3.4	3.5	0.6	2.6	−0.5

Source: World Bank (2005) World Development Indicators (averages calculated from annual data in the original source).

There are particular issues which need to be considered in making estimates of 'real' savings and investments in developing countries ('real' meaning values in terms of goods and services which may not enter within monetary market transactions rather than purely monetary values), and in comparing the nature of investment in developing countries with investment in developed market economics, some of which are elaborated in Box 4.2. In addition, it should be emphasised that a considerable proportion of investment in agriculture is often of a comparatively modest nature, not involving significant savings *per se* and monetary expenditure, so that the incremental capital–output ratio is likely to be low, as illustrated by Figure 4.2.

Inwards transfers of resources from abroad (through borrowing or through foreign direct investment) are a temporary expedient which can increase the funds available for productive investment. In the case of foreign borrowing, the indebtedness incurred needs to be serviced through interest payments and repayment of loans (which both

4.2 *Capital formation and innovation in agriculture*

The Kitinda Dairy Farmers' Cooperative in Western Kenya developed from a mid-1950s cattle dip (shown in this 1987 photograph) to the setting up of a modern small-scale milk processing plant under a Finnish aid project in the 1980s. This photograph demonstrates that significant investment in agriculture may be modest in value, can involve non-monetary labour inputs and technological change, and the benefits may substantially exceed the direct costs.

Photograph: Michael Tribe.

represent outflows from the economy). In the case of inflows of foreign direct investment, the outflows from the economy will arise because of the remission of profits and possibly capital repatriation as well (Box 4.2). Some discussion about the role of foreign aid in the economic growth of developing countries will be found in Chapter 8 of this book.

Box 4.2

The nature of capital formation and national income statistics in developing countries

Estimates for savings and investment in developing countries are usually based on the same system of national income accounts that is used for developed market economies. This is the United Nations System of National Accounts (UN, 1993). However, there are several characteristics of the economies of many developing countries which differ in significance as compared with developed market economies. One of these is non-monetary economic activity. In many developing countries, a considerable amount of investment still takes place outside the monetary economy – this particularly includes rural activities such as farm improvements, house construction and maintenance. The implications of this vary considerably between countries depending upon locally adopted national income accounting conventions. However, this factor could account for very low values of capital–output ratios for agricultural production in countries where agricultural investment expenditure is under-recorded (see Figure 4.2).

In an unpublished 1997 conference paper, Tribe compared the national income accounts approach to non-monetary elements in national income estimates (Tribe, 1997). Official estimates of non-monetary national income for Uganda show that non-monetary national income amounted to about 35 per cent of total GDP in 1975 and about 27 per cent of total GDP in 1995. By comparison, in Ghana no distinction is made between monetary and non-monetary national income in the published GDP statistics so that the consolidated data includes both elements (Republic of Ghana, 1996). This demonstrates that even though the basic national income accounting concepts tend to be 'universal' and based on the United Nations system, there is quite considerable variety in the application of the basic principles.

4.3 'New' theories of economic growth: technology, labour force enskillment and institutional change

Many of the concepts which are conventionally applied to the analysis of the contribution of capital accumulation to economic growth can also be applied to the analysis of the contribution of the labour force. However, it must not be forgotten that in the case of the labour force we are dealing with people rather than with the inanimate objects which comprise fixed capital formation.

If the age distribution of the labour force, and the participation rates for the various 'segments' of the labour force, remain constant, then the labour force will grow at the same rate as the population as a whole. If the labour force is growing at 3 per cent per annum and the capital stock is growing at the same rate, then the labour–capital ratio (which is more usually referred to as the capital–labour ratio) will remain constant. If the capital–labour ratio remains constant, economists will often think of this as representing a constant 'level' of technology and of labour productivity – and investment is referred to as 'capital-widening'. If the capital stock increases faster than the labour force, then each worker, on average, will have more capital to work with, the capital–labour ratio will rise, the 'level' of technology and labour productivity will rise, and investment is referred to as 'capital-deepening'.

Labour force 'participation rates' have been significantly affected by the proportion of women who are productively employed – a ratio which varies considerably between countries at different levels of development and with different socio-cultural traditions, as well as changing over time. The proportion of young people who are productively employed also tends to vary significantly between countries and over time – the higher the proportion of working-age children continuing in full-time education the lower the proportion who will be productively employed. In developed countries, the conventional age limits of the working-age population are 15 and 65 years (World Bank, 2005). Note that this definition does not necessarily involve any value judgements, but simply relates to a rigorous definition of population and labour force characteristics. Of course, in many developing countries young children (under 15) contribute productively to the economy through cattle-keeping, wood-gathering and water collection, releasing older members of the community for more highly productive activities. The same remark applies to the elderly (over 65s), who make a significant economic contribution through childcare and other household activity which, again, releases other more able-bodied members of the household for more productive work. This approach illustrates the economic concept of 'opportunity cost' – if the elderly did not undertake childcare the opportunity cost would be the output sacrificed through the loss of the activity of the more able-bodied who would not be released.

This discussion takes us into issues relating to the relationship between economic growth and technological change. There are several dimensions of technological change which are of economic significance, including 'consumption technology'. 'Consumption technology' refers to, for example, the inter-relationship between the characteristics of products and consumers. One obvious example of this is the continuum between the use of fresh food produce by consumers (either self-produced or purchased from the market), the use of preserved foods (dried, canned or frozen for example), and the use of prepared meals (either cook-chilled or frozen). Whichever consumption technology is employed implies the utilisation of different types of household equipment, so that the technological characteristics of products is economically significant. However, most of the concern with technology within the context of economic growth is with production technology and with associated technological – or technical – change.

Production technology is 'embodied' in the capital stock. At any given time the existing capital stock embodies a particular set of production technologies. A change of production technology requires changes in the characteristics of the capital stock and the rate of change of production technology is limited by the rate of change of the capital stock. Over time the rate of change of the capital stock is limited by: a) the rate of net investment – additions to the capital stock; and b) the rate of replacement investment – the replacement of old assets with new assets. This means that the rate of technological progress in the economy is closely, and positively, associated with the rate of gross fixed capital formation (GFCF).

Technology is also embodied in the labour force. The stock of skills and experience of the labour force at any one time is related to the existing types of technology, and the adoption of new production technology in the capital stock requires changes in the associated labour force skills. The characteristics of the labour force will therefore limit the level and rate of change of labour productivity if there is a mis-match with the characteristics of the capital stock. Higher levels of labour productivity require not only changes to the technological characteristics of the capital stock, but also changes to the 'technological characteristics' of the labour force – through education, training and enskillment in general.

The rate of technological change in the economy can be constrained by a) low levels of capital investment and by b) low levels of enskillment of the labour force. In turn, the rate of enskillment of the labour force is limited by two factors: a) the rate at which people with older skills leave the labour force (retirement), and the rate at which people with new skills join the labour force; and b) the rate at which the existing skills of the current labour force are upgraded through education and training – including vocational training.

Some of these labour market concepts have to be applied to developing countries with care. As compared with developed industrial countries, many of the developing countries have a high proportion of the labour force in 'unwaged' categories. Self-employment and family labour is considerably more significant in small-scale rural and urban enterprises than in the case in developed countries. In addition 'casually employed' labour is more significant in developing country labour markets. These features of labour markets are particularly important in interpreting comparisons of unemployment rates. However, in addition, in many developing countries the age limits between which people are regarded as being of 'working age' have little relevance, especially in rural and informal urban economic activity – with many people being economically active both below the age of 15 or 16 years, and above the age of 65 years.

4.4 Partial and total factor productivity: necessary but not sufficient conditions

It is possible to show some of these relationships in a relatively simple diagram which is based on neo-classical economic theory, principally related to the 'production function'. There has been quite significant criticism of the methodological foundations for the production function from heterodox economists, but it is widely used in economic analysis. The production function can be expressed verbally, algebraically or diagrammatically, and refers to the relationship between inputs and outputs in the production process. Figure 4.3 shows two production functions, Q_1 and Q_2, which relate to different time periods, Year 1 and Year 2. Conventionally total production (which is equivalent to total income and to total expenditure) can be referred to as Y (signifying income) or as Q (signifying production or output). This approach can be found in a number of publications, including Nishimizu and Robinson (1986), Nishimizu and Page (1988) and Syrquin (1986).

Starting in Year 1 we have the first of the production functions ($Q_1 = f_1$ (K,L)) showing that output (Q_1) in that year depends upon the amount of inputs employed (K,L) and upon the technical relationship between inputs and outputs (f_1). The technical relationship (f_1) can be explained as depending upon technology, the organisation of production and the skills of the labour force. Labour productivity, which might be thought of as depending largely upon the skills and efforts of the labour force, is very dependent upon the technology employed (in turn depending upon the size of the capital stock and the rate at which it is renewed), the efficiency of management of the production process, and institutional change. This is the reason for referring to this 'growth accounting' as *total* factor productivity.

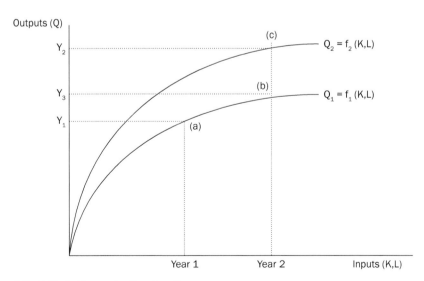

4.3 Shifts in the production function

Source: Nishimizu and Page (1988)

In Year 2, which may be several years after Year 1, the quantity of capital and labour available has increased and the production function ($Q_2 = f_2$ (K,L)) reflects the change in the technical relationship between inputs and outputs indicated by the switch from f_1 to f_2. The increased efficiency of production (referred to as total factor productivity growth – TFP growth) arises due to changes in the technology embodied in the capital stock, enskillment of the labour force, and improvements to the effectiveness of management. Other factors which can improve the efficiency of production over time include scale economies (lowering unit costs through the adoption of larger-scale production technologies) and 'external' economies (lowering unit costs through increased complementarity between producers). The TFP approach follows the contributions of Robert Solow to the theory of economic growth, on the basis of which empirical estimations of the 'residual' element of economic growth which cannot be accounted for by changes in the quantity of capital and labour inputs are made (see Box 4.3).

The movement from point (a) on the Year 1 production function to point (c) on the Year 2 production function shows economic growth. Part of this growth is due to an increased level of inputs (the movement along the horizontal axis from Year 1 to Year 2 and the associated movement along the Year 1 production function from (a) to (b)). Another part of the growth is due to total factor productivity (TFP) growth represented by the shift of the production function from ($Q_1 = f_1$ (K,L) to $Q_2 = f_2$ (K,L) and the movement from (b) to (c)).

It is then possible, in principle, to distinguish between changes in production levels (economic growth) which are based on:

a) changes in the level of inputs (movements along the production function);
b) changes in the nature of the production function (shifts of the production function).

More on technological change

If technological change is so important for economic growth, then the factors which affect the 'production of technology' are significant and it is necessary to understand the institutional and socio-economic

Box 4.3

Controversies around economic growth theory

A few years ago a journal article based largely on World Bank sources concluded that the differences between the economic growth experience of individual countries depended to a large extent on 'luck' (Easterly et al., 1993). It is necessary to distinguish between factors accounting for the *level* of income (which depend on the nature and extent of natural resource endowment, the size of the capital stock and the skills embodied in the labour force amongst other factors) and factors accounting for the *rate of growth* of income (which depend on changes in the international terms of trade, the rate of savings and investment, and the management of the economy amongst other factors). The article particularly focused on shocks as an explanation for cross-country variations in economic growth and on technological change as an explanation for long-run economic growth.

This raises the question of the extent to which external intervention – for example, through international aid and 'conditionality' – has a long-run influence on economic growth. This has been the subject of much research and writing in recent years (see, for example, McGillivray et al., 2006).

The theory of economic growth developed by Robert Solow is based on a neo-classical production function and relates the quantity of inputs to the quantity of outputs. However, this approach does not – strictly speaking – allow for changes in the long-run *quality* of inputs, or for changes in technology or in the institutional framework. This led to references to the *residual* element in economic growth – that part which cannot be accounted for by changes in the quantity of inputs. In other words, the *residual* explains total factor productivity (TFP) growth (see Figure 4.3). Some economists regard the *residual* as being very important in the context of economic growth theory (see, for example, Rosenberg, 2004; European Commission, 2001: 29, Appendix II.1) if only because depending upon which studies are being referred to it accounts for something between 0 per cent and 85 per cent of total economic growth. It can be understood why the *residual* is so significant. On the other hand, some economists who specialise in the study of economic growth theory do not specifically mention the *residual* at all (Weil, 2005).

circumstances relating to the generation of technological change. Technological change is of little economic significance unless it is actually adopted or utilized, meaning that the diffusion and adoption of technological change – the process through which it is incorporated into production – is of great interest.

One of the most prominent economists who wrote about this process was Joseph Schumpeter, who emphasised that the people who adopted changes – both technological and non-technological innovations – are risk-takers (see in particular Schumpeter, 1961). He described them as 'entrepreneurs', and described the changes as 'new combinations' which broke the 'circular flow' of production. In colloquial terms, the word 'entrepreneur' is often used interchangeably with 'businessman'; however, Schumpeter used the word more technically in order to refer specifically to entrepreneurship as a quality which includes significant elements of innovatory behaviour involving risk. The 'new combinations' which he specified included: a) production of new types of goods, or significant changes in the characteristics of existing goods; b) introduction of new methods of production, which may be based on new scientific discoveries; c) the opening up of a new market (either geographically or in a socio-economic sense within an existing location; d) the exploitation of new sources of raw materials or intermediate goods; e) the adoption of new systems of production (Schumpeter, 1961: 65–6). This analysis of economic growth and development still has considerable power within the narrative (as opposed to the quantitative or mathematical) tradition of economic theory. Schumpeter's approach to the determinants of economic growth raises questions about the significance of non-economic factors, including McClelland's 'need for achievement' and the Weberian 'Protestant ethic', which emphasise socio-psychological features of society (McClelland, 1967; Weber, 1930). It is clear that economic growth is a multi-dimensional phenomenon for which multi-disciplinary study is likely to yield more substantial results than single disciplinary approaches.

In recent years, economists have increasingly referred to 'endogenous growth', meaning that technological change is included *within* the economic growth model rather than being *outside* it (i.e. exogenous) (Romer, 1994; Pack, 1994). 'Endogenous growth' is also referred to as 'new' or 'modern' growth theory (Fine, 2006a; Thirlwall, 2006: Chapters 4 and 6; Todaro and Smith, 2008: 129 and Chapter 3, Appendix 3.3). The main significance of the word 'endogenous' is that Solow's growth model essentially treated technological change as being external (or exogenous), while empirical research has generally found that technological change accounts for a high proportion of economic growth. Having one of the main factors accounting for economic growth as 'exogenous' could be regarded as a major weakness of a growth model, hence the enthusiasm with which 'endogenous growth' has been greeted. However, we also know that institutional change and 'qualitative' changes (such as a better educated labour force) are also key factors accounting for economic growth, which are subject to influence by policy measures, rather than being 'residuals' which appear exogenously from outside the economic system. Although discussion about 'endogenous growth' may sometimes occur in an unsystematic manner, in a more political context it is encouraging that policy makers endeavour to understand the complexity of the factors which underlie economic growth and development.

4.5 Summary

We have an approach to economic growth which suggests that the following factors are important influences:

- the level of savings and investment
- the rate of change of the capital stock
- the rate of technical change
- the rate of growth of the labour force
- the rate of enskillment of the labour force.

The rate and type of institutional change also affects economic growth.

Questions for discussion

1 How important do you consider economic growth to be as an objective of societies and of governments' policies, particularly taking into account poverty reduction and environmental concerns in the early twenty-first century?

2 To what extent does economic growth depend upon significant savings and investment?

3 What is the role of technological progress in economic growth, and how is it included in the growth process?

4 How would you explain the differing economic growth experiences in recent decades between the countries of South East and East Asia and countries in sub-Saharan Africa?

5 How much does economic growth depend upon the luck of natural resource endowment and how much does it depend on good governance and public policy management?

Suggested further reading

Commission on Growth and Development. 2008. *The Growth Report: Strategies for Sustained Growth and Inclusive Development*. Washington: World Bank (for the Commission on Growth and Development) – downloadable from the Commission on Growth and Development's website.

Kenny, C. and Williams, D. 2001. What Do We Know About Economic Growth? Or, Why Don't We Know Very Much? *World Development*. 29 (1): 1–22.

Pack, H. 1994. Endogenous Growth Theory: Intellectual Appeal and Empirical Shortcomings. *Journal of Economic Perspectives*. 8 (1): 55–72.

Stern, N. 1991. The Determinants of Growth. *Economic Journal*. 101 (404): 122–33.

Todaro, M.P. and Smith, S.C. 2008. *Economic Development* (10th edn). Harlow: Addison Wesley Pearson. Chapter 3: Classic Theories of Economic Growth and Development.

Economic concepts used in this chapter

1 National income accounting:
Consumption
Savings
Investment; replacement investment (depreciation); net investment
Gross fixed capital formation
Gross domestic product; net domestic product
Capital accumulation   capital stock   working capital
Capital–output ratio   incremental   average
Disposable income

2 Economic growth:
Economic growth
Endogenous growth
Entrepreneurship   Innovations and 'new combinations'
Technology; production technology; consumption technology

3 Economic management:
   Economic management
   Economic fluctuations
   Neo-classical economic theory
   Capacity utilisation
   Productivity; total factor productivity
   Labour force; participation rates
   Production function

5 Economic growth and economic development since 1960

5.1 Introduction

The aim of this chapter is to relate the discussion of Chapters 2, 3, 4 and 6 to evidence of economic growth and development in developing countries since 1960. One of the main concerns is to establish the extent to which developing countries have experienced economic growth, and have 'developed' in an economic, as distinct from a non-economic, sense, over this period.

The impression is often given within the media that developing countries combine significant levels of poverty, conflict and corruption with negligible economic prospects, and with significant demands on higher-income developed countries for support from aid programmes. However, the evidence which is available in the form of economic statistics gives a much more positive picture, but there is considerable diversity. Much of the data which have been assembled for this chapter relate to global geographical regions – and even at this 'macro' level diversity is an obvious feature. Some of the data relate to individual countries because comparable data are not available on a wider regional basis. The reason for this is simply that data for regional groupings are the sum of the data for individual countries, so that if data for one country within a regional group is missing, then this means that the regional data will not be available. In Chapter 3, focusing on 'structural economic change', grouped data have been presented according to the World Bank's 'income group' categorisation, while in this chapter the regional groupings have been adopted. This gives a degree of variety to the book – and is also intended to emphasise that statistics need to be presented in different ways depending upon the particular issue which is the focus of the analysis, or which is being highlighted in the discussion.

When this book was being planned, the intention had been to discuss the experience of economic development since the end of the Second World War in 1945 in this chapter. This is a period which includes most of the decolonisation process. One of the implications of this was an increase in the extent to which the newly independent former colonies gained direct access to membership of a range of international institutions from which they had previously been excluded. The international 'scene' changed irrevocably, with many economic ramifications. However, it later became clear that the original intentions were difficult to achieve within the confines of the available space in the book, and so it has been necessary to limit discussion to a narrower range of issues.

One of the limiting factors for discussion of economic growth since 1945 is that in the earlier part of the period reliable and relevant statistics are very difficult to obtain, and those which are available for developing countries are often not directly comparable between individual countries, between groups of developing countries, or with statistical series for developed industrial countries. In recent years, the World Bank's 'World Development Indicators' (WDI) (World Bank, 2007a) have become available, including a wide range of statistical series for a large number of countries in a series which started in 1960. Parts of the WDI are readily available from the internet (i.e. the world wide web) and the full version is available from many libraries in a printed version or through the internet.[1] As a result of all of these limitations, particularly relating to the availability of comparable economic statistics, the discussion in this chapter has been restricted to the period between 1960 and 2005.

5.2 Economic growth

The World Bank's WDI (World Bank, 2007a) provide a growing range of statistics for 152 countries from 1960, and a more restricted range of statistics for another fifty-six countries covering the same period. There are detailed descriptions and notes which give significant information about the nature of the data (World Bank, 2007a). The WDI provide a basic statistical source which is widely regarded as being invaluable. The original statistics are provided by the national statistical offices of individual countries, but some of the national definitions are not completely consistent with the international definitions used for the WDI. For this reason, the World Bank statisticians undertake careful analysis and adjustment of the original statistics in order to ensure comparability of the published series and consistency with the definitions used for the WDI. Because of the large number of countries, statistical series and years involved, it is necessary to present a careful selection here as a basis for discussion of the economic development experience over the last four or five decades. The tables in this chapter refer to geographical regions, within which there has been considerable diversity of individual country experience over the period covered.

The first issue to be addressed is the economic growth of developing countries, comparing this evidence with equivalent data for developed industrial countries, and considering the relative economic growth performance of the different economic groupings shown in Figures 5.1 and 5.2 (based on the data in Table A 5.1 in the statistical annex to this chapter).[2]

Figure 5.1 clearly shows that the total GDP of the High-Income OECD countries dominates the international economy, with the total GDP of the European Monetary Union also being very significant by comparison with the regions of the world which include most of the developing countries. The higher-income economies have clearly also been growing at a sustained rate which has increased the 'income gap' relative to the developing regions of the world. Over the forty-five-year period from 1960 to 2005 the total GDP of the High-Income OECD countries grew by a factor of 4.5, while the East Asian and Pacific group of countries experienced twenty-fold GDP growth – a truly remarkable performance. However, by the end of the period the total GDP of the High-Income OECD countries was still over ten times the size of total GDP for the East Asian and Pacific group (compared with a factor of 47 in 1960). Table A 5.1 shows that the economic growth rate has consistently been between 5 and 10 per cent per annum for this group. By comparison the Latin American and Caribbean group had only a five-fold economic growth with a growth rate which faltered in the 1980s. The South Asian countries' growth was eight-fold, and sub-Saharan African growth was more than four-fold. The selected representatives of the developed industrial countries are the European Monetary Union and High-Income OECD countries,

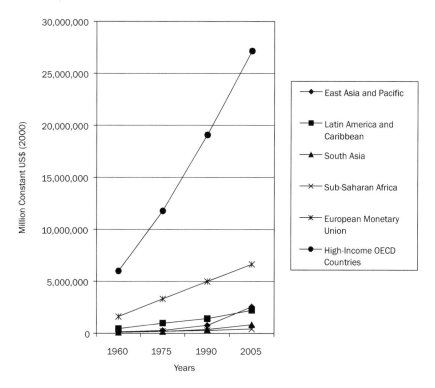

5.1 *Gross domestic product by country grouping (constant US$ 2000)*

Source: World Bank (2007a)

which experienced GDP growth of the same order as that for sub-Saharan Africa – i.e. towards the bottom end of the range – but, of course, from a starting point which was very considerably higher. The GDP growth data in Table A 5.1 show that most of the country groupings experienced a decline in the economic growth rate in the 1970s, 1980s and 1990s with higher growth rates at the beginning and end of this forty-five-year period.

The overall growth demonstrated in Figure 5.1 (and in the statistical annex tables) will confirm some preconceptions and contradict others. The exceptionally strong growth of the East Asian and Pacific group is largely accounted for by the strong economic performance of China and will probably come as no surprise – in 2005 China accounted for about 74 per cent of the GDP of this group. However, the fact that the sub-Saharan African group as a whole experienced quite substantial growth will perhaps be more unexpected, with weak growth in the 1970s and 1980s offset by much stronger growth in more recent years.

The South African economy is particularly important, accounting for as much as 38 per cent of all sub-Saharan GDP in 2005. This is an example of a common problem associated with the interpretation of regional economic statistics, where one country dominates the statistics for a regional grouping. The statistics themselves are not misleading, but the user of the statistics needs to exercise care in order to avoid drawing misleading conclusions.

Table A 5.1 also includes the levels and growth of per capita income, and it is here that the much lower absolute levels of GDP for East Asia and the Pacific, South Asia and sub-Saharan Africa in the 1960s become clearer (as is also demonstrated in Figure 5.2). The combination of differential GDP and population growth rates means that by the beginning of the twenty-first century East Asia and the Pacific had attained a level of per capita income (at constant prices for the year 2000 in US$) substantially more than double that of South Asia and sub-Saharan Africa from a starting point in the 1960s, which was about 20 per cent lower than that for East Asia and the Pacific and about 60 per cent lower than that for sub-Saharan Africa. It is customary to use constant prices (i.e. stripping out the influence of inflation) for these comparisons, with 'international US dollars' being the common currency adopted and the basis for these calculations is included in the notes and definitions for the World Development Indicators (World Bank, 2007a).

The data for per capita income growth over the period 1960 to 2005 in Figure 5.2 show that, despite fairly strong GDP growth, some of the

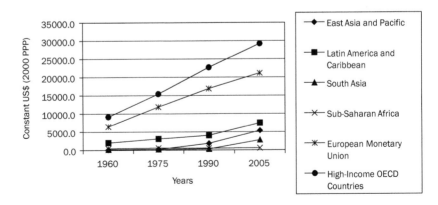

5.2 *Gross domestic product per capita by country grouping (constant US$ 2000)*

Source: World Bank (2007a)

developing countries have experienced a deterioration of their economic position compared with the developed industrial countries. Others have improved their relative position, but are still a long way behind the higher-income countries. In 1960, sub-Saharan Africa had a per capita GDP which was 4.7 per cent of that for the High-Income OECD countries, but by 2005 this had fallen to 2.5 per cent. For South Asia, the comparable statistics are 2.0 per cent and 2.5 per cent. On the other hand, for East Asia and the Pacific the comparable statistics are 1.5 per cent and 4.6 per cent. The sub-Saharan Africa statistics show the clear influence of comparatively weak GDP growth and comparatively high population growth.

Figure 5.2 and Table A 5.1 demonstrate that per capita income growth has been substantial in most regions of the world over the decades since 1960, but that the experience has been mixed. Over the entire period, East Asia and the Pacific recorded a ten-fold growth, as compared with a doubling in the Latin American and Caribbean region, a trebling in the South Asian region, and an increase of just over 30 per cent in the sub-Saharan African region. This compares with a trebling of per capita income in both the European Monetary Union and the High-Income OECD countries.

Table A 5.1 also contains data on constant PPP measures of GDP per capita from 1975. PPP stands for 'purchasing power parity' and allows for the fact that when national currencies are converted into US dollar values at the official exchange rate, the purchasing power of US dollars may be misrepresented. Put simply this is based on the fact that international exchange rates apply to internationally traded goods and services which typically have higher prices than domestically traded services and foodstuffs in most developing countries. The implication is that the 'constant 2000 US$' values of per capita income for most developing countries are relatively understated in terms of purchasing power, and those for developed industrial countries are relatively overstated. For sub-Saharan Africa, the PPP value of per capita income is about three times the 'uncorrected' US$ values. The difference between the two measures for East Asia and the Pacific is almost a factor of four, for Latin America and the Caribbean a factor of about 1.75 applies, and for South Asia a factor of well over four applies. It is notable that the PPP values of per capita income for the European Monetary Union and for the High-Income OECD countries are also higher than those for constant 2000 US$, but by a factor of only about 20 and 5 per cent respectively.

The PPP 'correction' demonstrates the 'sub-Saharan Africa' problem graphically. East Asia and the Pacific experienced a more than six-fold increase in the PPP per capita income measure between 1975 and 2005, as compared with a 33 per cent increase for Latin America and the Caribbean, an increase by a factor of 2.6 for South Asia, and a decrease of about 6 per cent for sub-Saharan Africa. It is this comparison which provided the basic argument for the establishment of the 'Commission for Africa' in 2004 (Commission for Africa, 2005). The contrast is made even clearer by the PPP measure of per capita income growth between 1975 and 2005 for the European Monetary Union, which increased by a factor of about 1.8 times, and for the High-Income OECD countries, which increased by a factor of about 1.9 times.

Nothing in this discussion of the PPP values for per capita income is intended to belittle the gulf which exists between the average levels of per capita incomes in developing countries and in the developed industrial countries. There is also no intention to belittle the significance of high levels of poverty in the developing countries (see Chapter 9). The PPP issue is only one of the serious limitations of the per capita GDP measure as an indicator of levels of living, as is explained in Box 5.1. However deficient the per capita income measure, it is still a clearly understood indicator of relative levels of income between countries and over time, and it is incontrovertible that higher standards of living associated with improved health conditions, education and water supply can only be achieved through expenditures (either public or private) based on higher levels of per capita income.

Another issue is the role of non-monetary, or non-marketed, production. For Uganda, the 2008 Statistical Abstract shows that over the period 2003 to 2007, 13 to 15 per cent of GDP was contributed by non-monetary economic activity. For agriculture, non-monetary value added was about 40 per cent of the total, and imputed income from owner-occupation of dwellings (which includes all types of dwellings in all parts of the country) accounted for about one-third of all non-monetary GDP (Republic of Uganda, 2008: 164). The Tanzanian national income statistics show that for 2002, 6 per cent of capital formation was non-monetary (National Bureau of Statistics Tanzania, 2002: Table F10). However, for Ghana the national income statisticians have aimed to consolidate the non-monetary categories into the main estimates for national income without distinguishing

Box 5.1

Limitations of the 'per capita income' measure

When asked to give an indication of the relative standard of living in different countries, or groups of countries, economists – and others – often give per capita income statistics in response. It is well known that this measure has serious limitations, but it also has one major advantage – giving a clear idea of the relative economic levels of different countries.

There are two groups of limitations – one related to shortcomings as an indicator of relative economic levels, and the other to shortcomings as a measure of relative standards of living.

The term 'per capita income' begs the question of which definition of 'income' is being used, and also of whether per capita consumption might be a better indicator of what is being sought. The gross domestic product (GDP) refers to the total value added generated within the boundaries of a country and is the most frequently used measure of 'income'. The gross national income (GNI) refers to total incomes within these boundaries and is equivalent to GDP less 'factor income sent abroad' plus 'factor income received from abroad'. In essence, one country's GNI is equivalent to GDP less profits, interest payments and interest paid to residents of other countries – remittances sent abroad might also be deducted – and plus profits, interest payments and interest (and possibly remittances) received from residents of other countries. Countries with large international debts, with substantial foreign investment within their boundaries, or sending out substantial levels of remittances will have a higher GDP than GDI, and vice versa. The third measure of 'income' is gross domestic expenditure, which is simply the total amount of expenditure within a country's boundaries – and which would usually be approximately the same as gross domestic income. The definitions for various forms of 'national income' can be found in the United Nations' 'System of National Accounts' (United Nations, 1994).

Another economic limitation of cross-country comparisons of per capita income measures relates to the fact that they necessarily involve conversion of values in national currencies to an international standard currency at the ruling official foreign exchange rate. These exchange rates only relate to the relative values of goods and services which are actually traded between countries, while goods and services which are not traded (e.g. a wide range of household and other services, many foodstuffs, and housing) may be relatively substantially much cheaper in some countries than in others. This means that international comparisons of per capita income should use the PPP (purchasing power parity values) referred to above in order to try to compensate for this problem.

continued

A third economic problem is that the average level of per capita income does not give any indication of the impact of income distribution (i.e. the differential extent of inequality).

A fourth economic problem is that in some countries a substantial amount of goods and services are not marketed and are therefore difficult to value in a way which can be included in the national income statistics. For example, in some countries a large amount of explicit employment and income is generated by household services while in others most of these services are provided on a non-marketed basis. Again, in some countries much investment in agriculture and in private housing is undertaken in a non-monetary way, and a considerable amount of foodstuffs consumed by much of the population is not marketed. Collectively this is referred to as 'subsistence' production, or non-monetary output, and some countries record it more systematically than others making comparisons difficult.

A fifth economic problem is that in many developed industrial countries a substantial amount of household income is spent on travelling to work, while this is less important in many developing countries. This issue is particularly related to the degree of urbanisation of the population. An associated issue relates to the proportion of household incomes devoted to the costs of heating (in cold locations) or of cooling (in hot locations).

In addition to these directly economic questions concerning the per capita income measure, there are many problems associated with the link between per capita income and 'welfare', wellbeing or the standard of living. Income measures of wellbeing have serious limitations many of which have been addressed by the development of 'levels of living indicators' such as the Human Development Index and the Human Poverty Index which are discussed in more detail in Chapter 9.

between monetary and non-monetary incomes (Republic of Ghana, 1996). This demonstrates that the statisticians in different countries have adopted various approaches to this issue, making international comparisons difficult.

5.3 Economic development and change

Chapter 3, focusing on structural change in the economy, concentrated on changes in production structures in particular. Here the intention is to consider a few other economic characteristics which can be regarded as central to 'economic development' as distinct from 'economic growth'.

Table 5.1 *Gross fixed capital formation (% GDP)*

	1960	*1975*	*1990*	*2005*
Brazil	n.a.	24.4	20.7	18.3
Indonesia	n.a.	n.a.	28.3	22.0
Malaysia	11.9	25.6	33.0	20.0
Kenya	n.a.	20.2	20.6	18.6
South Africa	19.6	29.0	19.1	17.2
India	12.7	16.2	22.9	28.1
France	n.a.	23.7	21.6	19.7
United Kingdom	16.9	20.7	20.5	16.6
Japan	29.6	33.1	32.5	n.a.
United States	17.9	17.8	17.4	n.a.

Source: World Bank (2007a).

Table 5.1 shows data for gross fixed capital formation (GFCF) – in other words, investment – for a forty-five-year period from 1960. Following from the discussion in Chapter 4, the term GFCF refers to a) gross investment – that is, both new and replacement investment – and b) fixed investment – that is, including investment in capital goods, buildings and infrastructure (e.g. roads, harbours, railways, water, energy, etc.) and excluding changes in working capital. It can be seen that by the end of the period all of these countries were sustaining a rate of investment in the region of 20 to 25 per cent of GDP, although some had experienced fluctuations. This is consistent with the economic logic (or theory) outlined in Chapter 4. Several countries had achieved an upwards shift in the rate of investment, notably India and Malaysia, and these countries have been regarded as being among the best economic performers in recent years. The statistics for 1990 are historically high for Malaysia and Indonesia, while India has its highest ratio in 2005. Japan, long regarded as a comparator in terms of economic performance, has a noticeably high rate of investment throughout the period. The mature industrial economies – France, the United Kingdom and the United States – have rates of investment which are steady, but perhaps a little lower than might have been expected.

Table 5.2 presents statistics for the period from 1965 for what the WDI defines as 'General Government Final Consumption Expenditure'. Statistics for 1960 were not available in the World Development Indicators for sufficient countries, and so 1965 has been substituted as the starting year for this particular table. One conclusion, from the variations between the statistics between

Table 5.2 *General government final consumption expenditure (% of GDP)*

	1965*	1975	1990	2005
Brazil	14.1	10.6	19.3	19.5
Indonesia	10.4	9.0	8.8	8.2
Malaysia	10.6	17.1	13.8	12.9
Kenya	11.0	18.3	18.6	17.1
South Africa	9.7	14.8	19.7	20.2
India	7.0	9.9	11.6	11.3
France	16.5	19.2	21.7	23.7
United Kingdom	16.5	22.3	19.8	21.8
Japan	11.5	14.4	13.4	n.a.
United States	16.5	17.9	17.0	n.a.

Source: World Bank (2007a).

Note: * Data for 1965 have been given because there was insufficient data available for these countries for the year 1960.

countries and over time, is that governments in the more 'mature' developed industrial countries tend to command a higher proportion of final consumption expenditure than governments in developing countries. A cross-country comparison of Indonesia, India, France and the United Kingdom tends to support this interpretation, while a comparison of the statistics over time for Brazil and South Africa are also broadly supportive of this view. One of the reasons for the differences in the government's share of the GDP, and of consumption expenditure in particular, is that governments in the more-developed economies tend to have a much wider (and more elaborate) range of services which they provide, including enhanced environmental protection, health and safety regulations and competition policy for example. Another reason for expecting 'development' to lead to higher government involvement in the economy is that state social security and social work/care systems tend to become more elaborate in the long term – involving transfers from economically active to economically inactive members of communities, and – significantly – between different generations of the population.

Table 5.3 shows data from the WDI for 'Trade' over the period from 1960 to 2005, showing a dramatic increase in international openness for almost all of the countries included in the table. The statistical exceptions are Kenya, South Africa and Japan, although even these three economies will also be more 'open' than at the start of the period if alternative measures had been used. There have been two associated

Table 5.3 *Trade* (% of GDP)*

	1960	1975	1990	2005
Brazil	14.2	19.0	15.2	29.2
Indonesia	26.9	45.0	49.1	62.8
Malaysia	89.0	85.6	147.0	223.3
Kenya	64.8	64.3	57.0	62.3
South Africa	56.0	57.8	43.0	55.7
India	11.9	12.9	15.7	44.7
France	26.7	36.7	44.0	53.1
United Kingdom	41.8	52.5	50.6	56.1
Japan	21.0	25.6	20.0	n.a.
United States	9.6	16.1	20.5	n.a.

Source: World Bank (2007a).

Note: * Trade is defined as merchandise imports plus merchandise exports.

forces which account for this increase in openness. The first is the general phenomenon of 'globalisation', which is discussed in more detail in Chapter 6. The second is the impact of regional economic groupings – the European Union in the case of France and the United Kingdom, and the North American Free Trade Area in the case of the United States. The increased importance of transnational corporations (TNCs) in the world economy has been linked to both of these economic forces, and the outcome has been increased international dependence on trade, reflected in these statistics.

5.4 Population growth

This book has not aimed to be comprehensive in its approach to all aspects of development economics and of developing economies. One of the issues which has not been highlighted is that of population and its interaction with economic development. However, we feel that two dimensions of 'population and development' cannot be omitted – the first is that of the extent to which differences in population growth rates between countries and over time explain differences in per capita income growth, and the second is the 'demographic transition'.

Part of the difference in the observed per capita income performance between the regions and countries included in the comparisons in the previous sections of this chapter is due to variations in economic

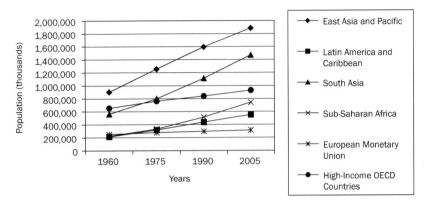

5.3 *Total population by country grouping 1960 to 2005*

Source: World Bank (2007a)

growth experience. The East Asia and Pacific region has had historically very high economic growth for most of the period since 1960. By comparison, the sub-Saharan African region's growth recovery in the latter part of the period cannot compensate for the 'lost' years during which the GDP growth rate was lower – low or negative economic growth over a substantial period means that the 'economic level' at the end of the period will be substantially lower than that for comparator countries. However, much of the difference in the observed per capita income performance of the economic groupings is accounted for by the variations in population growth – a simple arithmetic fact.

Figure 5.3 presents statistics for total population for the same regions and groupings used earlier in this chapter that are based on the information contained in Table A 5.2 in the statistical annex to this chapter. Over the period 1960 to 2005 world population approximately doubled. The population of East Asia and the Pacific region increased at about the same rate as this world figure. However, the population of sub-Saharan Africa increased by a factor of 3.3 in the same period, and that of Latin America and the Caribbean and of South Asia increased by a factor of about 2.5. The higher-income countries of the other two groups (the European Monetary Union and High-Income OECD countries) experienced population growth by a factor of around 1.3 (1.24 and 1.41 respectively).

Figure 5.4 shows that the rate of growth of population has been declining around the world, but that in 2005, at the end of the period,

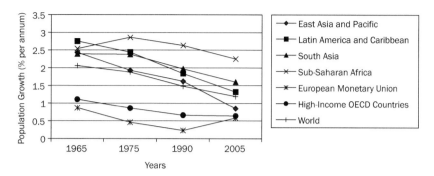

5.4 *Population growth rate by country grouping 1965 to 2005*

Source: World Bank (2007a)

Note: the WDI data for 1960 are seriously inconsistent and so the series has been started in 1965.

the population of sub-Saharan Africa was growing at about 3.75 times the rate of the two high-income groupings, and about 2.6 times the growth rate for East Asia and the Pacific. While sub-Saharan Africa had the highest growth rate throughout most of the period, it first rose from about 2.5 per cent per annum in 1965 to a peak of about 2.75 per cent in 1975 and then fell back to about 2.25 per cent in 2005. The growth rate for East Asia and the Pacific fell steadily throughout the period from about 2.4 per cent per annum in 1965 to about 0.8 per cent in 2005, while that for South Asia fell from about 2.3 per cent per annum in 1965 and 1975 to about 1.6 per cent in 2005. The substantial falls in the population growth rates for East Asia and the Pacific (statistically dominated by China) and for South Asia (statistically dominated by India) are both related to government policies intended to achieve this type of reduction. The growth rate for Latin America and the Caribbean fell from about 2.75 per cent per annum in 1965 to about 1.3 per cent over the same period, but this occurred without any particular set of government policies directed at population growth. This raises the question of whether there are 'endogenous' socio-economic forces which drive population dynamics, and whether population growth in South and East Asia would have fallen significantly even in the absence of government policies.

Figure 5.4 also shows that the population growth rate in the higher-income countries of the High-Income OECD group was below 1 per cent per annum for most of the period, and that for the European

Monetary Union (EMU) the growth rate fell from about 0.8 per cent per annum in 1965 to about 0.25 per cent in 1990 but then rose to about 0.6 per cent in 2005. Over the period, the world population growth rate has fallen consistently, from about 2.1 per cent in 1965 to about 1.2 per cent in 2005.

This population growth experience goes some way towards explaining the changes to per capita income growth which were discussed earlier in this chapter. Quite clearly, the higher the rate of population growth, the lower the rate of per capita income growth for any given rate of economic growth. This issue has already been discussed in Chapter 4 in the context of economic growth rates. Thus, although sub-Saharan Africa has experienced significant economic growth over the period covered by this chapter, per capita income growth has been low by comparison with other global regions. Because per capita income growth is one of the principal elements of a poverty reduction strategy (refer to Chapter 9), the higher the rate of population growth, the more difficult it is to achieve poverty reduction targets. Wider issues relating to population and international development are discussed in Todaro and Smith (2008: Chapter 6), in Thirlwall (2006: Chapter 8) and in the Commission on Growth and Development (2008: Statistical Appendix Section 2).

We turn now to the age distribution of the population – which is directly linked to the issue of the 'demographic transition'. The basis for the demographic transition is that, taking a broad historical view, at 'early stages of development' populations tend to have relatively high death rates, and – in a sense to compensate for this – relatively high birth rates. In more detail, at these 'early stages of development' there tends to be high infant and child mortality, high maternal mortality and low life expectancy. As living conditions have improved, with better water supplies and sanitation, better nutrition and better curative and preventive medical care, death rates have fallen significantly and life expectancy has risen significantly. With lower death rates and unchanged, relatively high, birth rates, the rate of population growth will increase – so that it is only with lower birth rates that the rate of population growth would fall back to the rates which had applied in the past. There are socio-economic forces which give an incentive for families to restrict the number of children as income levels rise and other socio-economic changes occur. There tends to be a 'natural' tendency for population growth rates to decline from the heights reached following the reductions in death rates which have been achieved in the last ten to fifteen decades. However, there are also strong pressures from governments, and internationally, for birth rates to be reduced faster than would be achieved by these 'natural' socio-economic trends.

Figure 5.5 is reproduced from an interesting undated document posted on the internet at the University of Wisconsin, Marathon County (Montgomery, 2009). In many cases, internet sources are of doubtful quality, but this source is of good quality. The figure shows the general case, where in Stage One there are high birth rates and high death rates, and a comparatively low population growth rate. In Stage Two the death rate has fallen, due to the impact of better health and nutrition, but the birth rate remains at the higher level from the previous stage. In Stage Three the death rate levels out at a historically much lower level and the birth rate falls substantially, so that the population growth rate falls back to a level much closer to that in Stage One. Finally, in Stage Four birth and death rates are at a sustained lower level with low population growth rates. This is an example of a model – not an economic model – which outlines demographic experience in the long-term, and which can be explained partially by economic factors, although social factors will be of considerable significance.

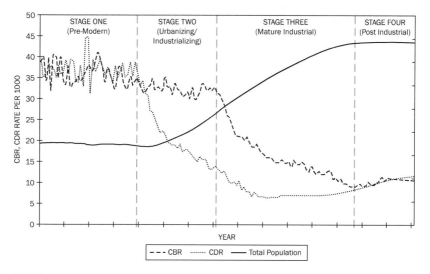

5.5 *The demographic transition*

Source: Montgomery (2009)

As the population growth rate in developed industrial countries has declined, and as the life expectancy has increased, the shape of the 'population pyramid' has changed so that a smaller proportion of the population are children and a higher proportion are older people. It is this combination of changes in population growth rates ('population dynamics') and in the age structure of the population which are important for the 'demographic transition'.

Between 1960 and 2005 the crude birth rate for East Asia and the Pacific fell from 26.5 to 14.7 per 1000 population, for Latin America and the Caribbean it fell from 41.2 to 20.5, for South Asia it fell from 47.4 to 24.7, while for the European Monetary Union it fell from 18.6 to 10.3 and for High-Income OECD countries it fell from 20.9 to 11.3. By comparison, the crude death rate for East Asia fell from 23.9 to 6.7, for Latin America and the Caribbean it fell from 13.0 to 6.0, for South Asia it fell from 23.5 to 7.7, while for the European Monetary Union it fell from 10.6 to 9.4 and for High-Income OECD countries it fell from 9.9 to 8.6 (data from World Bank, 2007a).

In sub-Saharan Africa, the crude birth rate fell from 48.4 per 1000 population in 1960 to 39.5 per 1000 in 2005, while the crude death rate fell from 24.0 per 1000 to 17.4 per 1000, with the result that the natural population growth rate has declined only slightly from the high rates experienced in the 1960s (from 2.4 per cent per annum to 2.2 per cent) (World Bank, 2007a). Some of these issues are discussed in more detail in Chapter 9 relating to poverty.[3]

The statistics presented in Figures 5.6 and 5.7 demonstrate the 'demographic transition' remarkably well, putting flesh on the general case shown in Figure 5.5. For the 'developing country' groupings (East Asia and the Pacific, Latin America and the Caribbean, South Asia and sub-Saharan Africa), the proportion of the population in the age group 0 to 14 years was about 40 per cent in 1960, while for the two higher-income groupings (the European Monetary Union and High-Income OECD countries) was around 25 to 28 per cent. By the end of the period, in 2005, the proportion in the younger age group had fallen to 24, 30 and 33 per cent for three of the developing regions, and remained at 43.5 per cent for sub-Saharan Africa. In the higher-income groupings, the proportion of population aged 0–14 had fallen to the 15.5 to 17.6 per cent range.

Turning to the higher-age group (those over the age of 65 years), shown in Figure 5.7, for the four 'developing country' groupings in

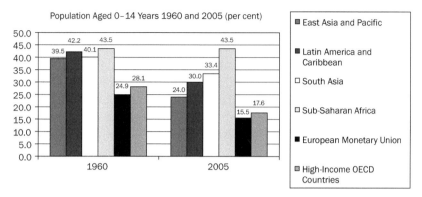

5.6 *Proportion of population aged 0–14 years*

Source: World Bank (2007a)

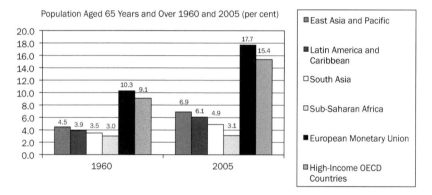

5.7 *Proportion of population aged 65 years and over*

Source: World Bank (2007a)

1960 the proportion was in the range of 3.0 to 4.5 per cent, while the equivalent proportions for the higher-income groupings were in the range of 9 to 10 per cent. By 2005, the proportion of the population in the older-age group had increased to 5 to 7 per cent for the first three of the developing country groupings and had risen from 3.0 to 3.1 per cent in sub-Saharan Africa. By comparison the higher-age group had increased to 17.7 and 15.4 per cent in the European Monetary Union and the High-Income OECD countries respectively.

One aspect of this 'demographic transition' is the relationship between the age structure of the population and what is referred to as the 'dependency ratio'. This dependency ratio is intended to capture the significance of the ratio of 'producers' to 'non-producers' in a society. Another way of presenting the same information would be to refer to the ratio of producers to consumers. The higher the ratio of consumers to producers, the greater the downwards pressure on a) per capita income, and b) on the savings ratio. Conventionally, in developed industrial countries which have widespread formal contracts of employment and pensions schemes, the 'working-age population' is defined as all males and females between the ages of 15 and 65 years. The labour force 'participation ratio' is defined as the ratio of those actually working (or 'employed') between these ages and the total number in these age groups. Obviously, the higher the proportion of the population below the age of 15 and above the age of 65, the smaller the proportion 'available for work' – with a lower balance of producers to consumers. The 'participation ratio' is affected by a range of social and economic factors including the position of women in society and the extent of involuntary unemployment.

In most developing countries, a high proportion of working people are not in formal contractual relationships with their employer, there are no unemployment benefit schemes, and no widespread formal pension schemes. Age is also not as important in determining whether people are making a 'productive' contribution to the economy. This means that although the 'age pyramid' is an important element of the analysis of the role of population growth in economic development, its role is somewhat different in developed industrial countries and in developing countries.

5.5 Summary

- The main aim of the chapter was to establish the extent to which developing countries have experienced economic growth and development since 1960, and the extent to which their comparative position relative to higher-income countries has changed since 1960.
- There has been considerable diversity of experience over the period, with very rapid economic growth in East Asia and the Pacific, and slower growth in sub-Saharan Africa at the extremes.
- Despite the rapid growth of the East Asian and Pacific economies, the gap between their income levels and those of the higher-income countries has increased in both absolute and per capita terms.

- Rapid population growth in sub-Saharan African countries has led to falling average per capita incomes despite positive economic growth.
- 'Purchasing power parity' measures of per capita income indicate that the 'gap' between developing and higher-income economies is smaller than that suggested by unadjusted data, but the difference is still very substantial.
- There have been significant changes in a number of indicators relating to economic development, including significant statistical evidence of increased 'openness' of economies.
- Developing countries, and sub-Saharan Africa in particular, have experienced historically high rates of population growth because death rates have fallen considerably faster than birth rates.
- East Asia and the Pacific and South Asia (specifically China and India) have experienced remarkable reductions in birth rates following robust government population policy, although Latin American birth rates have also fallen (but not so dramatically) without comparable government policy interventions.
- Global population data over the last four to five decades tend to confirm the 'demographic transition' – with population growth rates falling due to socio-economic factors independently of government policies.

Questions for discussion

1 Why, despite developing countries having relatively high economic growth rates, has the 'income gap' between developing and high-income countries become larger over the last four decades?

2 Discuss the range of possible reasons for the decline in per capita income in sub-Saharan Africa over the last four decades.

3 What have been the main features of global 'economic development' in the last four decades?

4 What is the 'demographic transition' and is there any evidence that it is occurring in developing countries?

Suggested further reading

Commission on Growth and Development. 2008. *The Growth Report: Strategies for Sustained Growth and Inclusive Development.* Washington: World Bank (for the Commission on Growth and Development).

Thirlwall, A. 2006. *Growth and Development with Special Reference to Developing Countries* (8th edn). Houndmills, Basingstoke: Palgrave Macmillan, Chapter 8.

Todaro, M. and Smith, S. 2008. *Economic Development* (10th edn). London: Pearson Addison Wesley, Chapter 6.

World Bank. 2007. *World Development Indicators 2007.* Washington, World Bank.

Economic concepts used in this chapter

1 Economic growth and development:
   Economic growth
   Economic development
   Structural change

2 National income accounting:
   UN system of national accounts
   Value added
   Gross domestic product
   Gross national income
   Gross domestic expenditure
   Gross fixed capital formation
   Gross investment
   Fixed investment (new and replacement investment)
   Working capital
   General government final consumption expenditure
   Economic 'openness'/trade
   Subsistence production (non-monetary/non-marketed output)
   Purchasing power parity (PPP)
   International exchange rates
   Internationally traded goods
   Domestically traded goods
   Constant prices

3 Standards of living:
   Welfare/wellbeing

Levels of living indicators
Per capita income

4 Population dynamics:
Demographic transition
Dependency ratio/Labour force participation ratio
Socio-economic forces/factors

Table A 5.1 (Statistical Annex) Economic growth between the early 1960s and 2005

		1960	1975	1990	2005
Total GDP (millions constant 2000 US$)	East Asia and Pacific	127,100	265,240	767,490	2,555,100
	Latin America and Caribbean	437,540	981,880	1,425,900	2,227,100
	South Asia	102,880	176,690	364,940	831,210
	Sub-Saharan Africa	97,697	194,700	273,000	423,020
	European Monetary Union	1,623,500	3,322,500	4,991,300	6,638,300
	High-Income OECD Countries	5,992,900	11,768,000	19,080,000	27,148,000
GDP growth (annual %) (a)	East Asia and Pacific	10.6	6.8	5.5	9.0
	Latin America and Caribbean	5.7	3.4	0.4	4.5
	South Asia	-0.6	7.0	5.6	8.7
	Sub-Saharan Africa	6.3	1.1	1.1	5.7
	European Monetary Union	4.5	-0.7	3.7	1.3
	High-Income OECD Countries	5.0	0.4	3.0	2.5
GDP per capita growth (annual %) (a)	East Asia and Pacific	7.9	4.8	3.9	8.0
	Latin America and Caribbean	2.9	0.9	-1.4	3.1
	South Asia	-3.0	4.5	3.4	6.9
	Sub-Saharan Africa	3.7	-1.7	-1.7	3.4
	European Monetary Union	1.4	-1.1	2.5	0.7
	High-Income OECD Countries	3.8	-0.5	2.2	1.9

Table A 5.1 (continued)

		1960	1975	1990	2005
GDP per capita (constant 2000 US$)	East Asia and Pacific	141	211	481	1,355
	Latin America and Caribbean	2,033	3,088	3,259	4,044
	South Asia	183	221	328	566
	Sub-Saharan Africa	434	587	531	569
	European Monetary Union	6,455	11,877	16,904	21,148
	High-Income OECD Countries	9,144	15,419	22,712	29,251
GDP per capita, PPP (constant 2000 international $)	East Asia and Pacific	n.a.	813	1,857	5,384
	Latin America and Caribbean	n.a.	5,639	6,035	7,482
	South Asia	n.a.	1,085	1,601	2,791
	Sub-Saharan Africa	n.a.	1,878	1,678	1,774
	European Monetary Union	n.a.	14,500	20,649	25,944
	High-Income OECD Countries	n.a.	15,796	23,068	30,058

Source: World Development Indicators (World Bank, 2007a).
Notes: n.a. = not available. (a) = Growth rate given for 1960 is that for 1960 to 1965.

Table A 5.2 (Statistical Annex) World population between 1960 and 2005

		1960	1975	1990	2005
Total population (thousands)	East Asia and Pacific	902,640	1,255,100	1,596,100	1,885,500
	Latin America and Caribbean	215,200	317,990	437,560	550,770
	South Asia	562,490	799,610	1,113,100	1,469,800
	Sub-Saharan Africa	224,950	331,660	514,360	743,060
	European Monetary Union	251,520	279,740	295,280	313,900
	High-Income OECD Countries	655,400	763,230	840,090	928,120
	World	3,026,500	4,061,300	5,256,300	6,437,700
Population growth (annual %) (a)	East Asia and Pacific	2.43	1.93	1.62	0.85
	Latin America and Caribbean	2.75	2.44	1.84	1.32
	South Asia	2.39	2.38	1.97	1.60
	Sub-Saharan Africa	2.54	2.86	2.63	2.25
	European Monetary Union	0.87	0.46	0.23	0.58
	High-Income OECD Countries	1.11	0.86	0.66	0.64
	World	1.19	1.88	1.48	1.19
Population ages 0–14 (% of total)	East Asia and Pacific	39.5	40.2	30.3	24.0
	Latin America and Caribbean	42.2	41.4	36.4	30.0
	South Asia	40.1	40.5	37.8	33.4
	Sub-Saharan Africa	43.5	45.2	45.6	43.5
	European Monetary Union	24.9	23.9	18.0	15.5
	High-Income OECD Countries	28.1	25.0	19.8	17.6
	World	36.8	36.7	32.5	28.1

Table A 5.2 (continued)

	1960	1975	1990	2005
Population ages 65 and above (% of total)				
East Asia and Pacific	4.5	4.2	5.1	6.9
Latin America and Caribbean	3.9	4.3	4.7	6.1
South Asia	3.5	3.8	4.1	4.9
Sub-Saharan Africa	3.0	2.9	2.9	3.1
European Monetary Union	10.3	12.8	14.4	17.7
High-Income OECD Countries	9.1	10.9	12.9	15.4
World	5.3	5.7	6.2	7.4
Crude birth rate (per 1000 population)				
East Asia and Pacific	26.5	n.a.	22.9	14.7
Latin America and Caribbean	41.2	n.a.	26.6	20.5
South Asia	47.4	n.a.	32.0	24.7
Sub-Saharan Africa	48.4	n.a.	44.9	39.5
European Monetary Union	18.6	n.a.	11.5	10.3
High-Income OECD Countries	20.9	15.0	13.5	11.3
World	32.8	n.a.	25.7	20.2
Crude death rate per 1000 population				
East Asia and Pacific	23.9	n.a.	7.1	6.7
Latin America and Caribbean	13.0	n.a.	6.8	6.0
South Asia	23.5	n.a.	10.3	7.7
Sub-Saharan Africa	24.0	n.a.	16.7	17.4
European Monetary Union	10.6	n.a.	9.9	9.4
High-Income OECD Countries	9.9	9.3	8.8	8.6
World	18.6	n.a.	9.3	8.7

Source: World Development Indicators (World Bank, 2007a).

Notes: n.a. = not available. (a) Note that for population growth the data in the 1960 column are for 1965 – the 1960 WDI statistics have inconsistencies for this series.

Notes

1 There are several other international statistical sources available including those from the OECD, the United Nations, the UNDP and regional bodies such as the African, Inter-American and Asian Development Banks. Another statistical source is the *Penn World Tables* from the University of Pennsylvania (Heston et al., 2006). A good overall gateway or portal into a wide range of information is provided from ELDIS at www.eldis.org.

2 The tables in the statistical annex at the end of the chapter provide the supporting date for the figures which have been included in the text.

3 A United Nations-sponsored resource for information about African Population is the *Regional and National Population Information – Africa* website, which can be accessed at www.un.org/popin/regional/africa/

Appendix: a case study of Ghana and Uganda

Ghana (1957) and Uganda (1962) are two sub-Saharan African countries which achieved political independence within ten and fifteen years respectively of the South Asian sea-change which saw the creation of India and Pakistan as independent states in 1947. Both of these African states, Ghana in the west and Uganda in the east, were regarded as having a very good educational base by comparison with other sub-Saharan African countries, and both were economically highly dependent on their agricultural sector in terms of national income and export revenue. Both countries experienced periods which included elements of economic stagnation and decline between the mid-1970s and the mid-1980s, following which there was quite strong recovery. In the case of Ghana, the decline came about mainly because of economic mismanagement by military governments but without significant civil unrest. In the case of Uganda, the decline was associated with a military government in the earlier period (1971–79), and in the later period there was a nominally civilian government in power but with significant internal conflict, which ended in 1986 when the National Resistance Movement achieved power. Sources for basic information about Ghanaian and Ugandan experiences over this period include Huq (1989), Baah-Nuakoh (1997), Aryeetey et al. (2000) and Reinikka and Collier (2001). This case study is intended to set out a few of the economic dimensions of these experiences. Some economic statistics have been obtained from the *World Development Indicators* (World Bank, 2007a), as well as from national statistical sources, which are presented in Tables A.1 and A.2.

In Table A.1, the statistics for gross domestic product (GDP) show that no data is available for Uganda up to 1980 from the *World Development Indicators*. This must demonstrate the problem of comparability between series of statistics based on different statistical concepts, because a considerable amount of economic data was published by the Ugandan government over the period 1960 to 1980. For Ghana, the GDP data show a period of stagnation between 1970 and 1985 following which fairly strong growth occurred. The per capita GDP data, including the influence of population growth, show a significant decline between 1970 and 1985, after which growth resumed so that by the year 2005 the level returned to that achieved in 1970. For Uganda, the GDP data in Table A.1 show steady growth from 1985, and the same applies to GDP per capita. By comparison, the GDP 'growth' data from national sources in Table A.2 for both Ghana and Uganda show negative values for seven of the years (Ghana) and for five of the years (Uganda). Referring to the basic data (which has not been included in the table) for Ghana shows that the level of GDP attained in 1974 was not re-attained until 1988, with the lowest value (for 1983) being only 78.5 per cent of the 1974 value. For Uganda, the 1974 level of GDP was not re-attained until 1987, with the lowest value (for 1980) being 81.5 per cent of the 1974 value. This represents a drop of 20 to 25 per cent in GDP, with 'recovery' and renewed growth being achieved in both cases following significant political and economic 're-structuring'.

A limited amount of data is shown in Table A.2 for 'inflation', based on national statistics. In both Ghana and Uganda, relatively high rates of inflation are shown for several years between 1974 and 1984, although in more recent years the rates of inflation have been considerably lower – representing part of the results from the economic recovery process.

Other statistics in Table A.1 show some interesting aspects of the economic stagnation/decline and recovery. For example, agricultural value added (its contribution to GDP) increased markedly over the period during which GDP stagnated or declined, suggesting that agriculture is still a 'mainstay' of the economy – with almost all food consumed within these countries being produced domestically. However, in more recent years, the proportion of agricultural value added in GDP has fallen back to levels below those of the 1960s. Manufacturing value added was a comparatively low proportion of GDP in 1980, and in the case of Uganda, these low levels continued until the mid-1990s.

From Table A.1 it can be seen that the period during which GDP stagnated or fell was associated with significant falls in exports relative to GDP. In the case of Uganda, the recovery has been considerably less strong than in Ghana, but Ghana has benefited from being able to export gold, while Uganda has had difficulty re-attaining its former international market position. Merchandise trade (i.e. imports plus exports) as a proportion of GDP confirms this 'difficult' trade position for Uganda.

The gross fixed capital formation (i.e. investment) data show that the economic recovery of the 1990s and 2000s has been associated with higher investment levels. However, this investment has not been financed from domestic economic sources, so that inflows from abroad have been important means of financing the higher investment levels. This issue represents a major concern for the future health of these economies.

What are the main conclusions arising from the economic experience represented by the data in the tables and other information contained in the publications cited at the beginning of this brief appendix? For low-income sub-Saharan African countries, the recovery from economic decline and stagnation demonstrates a degree of resilience which is perhaps surprising. Particularly notable is the recovery of the manufacturing sector which, for both countries, has been considerably based on the type of import-substituting industries frowned upon by economists inclined towards trade liberalisation and export promotion. Another notable feature of the economic performance during the 'difficult years' was the dependability of the agricultural sector, much of which is still based on smallholder production. Indeed, in many senses it could be said that over the last four decades there has been comparatively little change to the basic economic structures of both countries.

The recovery of both economies was related to considerable inflows of international aid, and this has been particularly important in the case of Uganda. In addition, Ghana has the advantage of a gold mining industry which undertook a major investment and rehabilitation programme following the initial economic recovery in the mid- to late 1980s, so that robust foreign exchange earnings were then earned from gold exports. It should be emphasised that following a decade of decline and another decade of recovery both Ghana and Uganda found themselves at about the same level of per capita income that they had achieved twenty years earlier. In this sense, both countries had sacrificed opportunities for substantial increases in living standards by comparison with countries which had sustained economic growth rather than periods of decline and recovery over this period.

Finally, one of the main conclusions must be that all countries have their own individual characteristics, so that factors leading to the decline and recovery of the economies are specific to these countries and generalisations have to be made with care.

Table A.1 A selection of Ghanaian and Ugandan economic statistics 1960 to 2005

		1960	1965	1970	1975	1980	1985	1990	1995	2000	2005
GDP (constant 2000 US$ million)	Ghana	1,897.9	2,210.7	2,546.9	2,515.6	2,637.0	2,580.9	3,263.4	4,024.6	4,972.1	6,357.2
	Uganda	..	..	..	..	..	2,404.3	3,076.7	4,317.3	5,926.4	7,786.2
GDP per capita (constant 2000 US$)	Ghana	266	270	284	246	233	193	211	227	250	287
	Uganda	..	..	..	..	..	163	173	207	244	270
Agriculture, value added (% of GDP)	Ghana	40.8	43.5	46.5	47.7	57.9	44.9	44.8	38.8	35.3	37.5
	Uganda	51.7	52.3	53.8	72.2	72.0	52.7	56.6	49.4	37.3	32.7
Manufacturing, value added (% of GDP)	Ghana	..	9.8	11.4	13.9	7.8	11.5	9.8	9.3	9.0	8.3
	Uganda	8.5	8.4	9.2	6.3	4.3	5.8	5.7	6.8	9.8	9.2
Exports of goods and services (% of GDP)	Ghana	28.2	17.1	21.3	19.4	8.5	10.7	16.9	24.5	49.1	36.1
	Uganda	27.3	25.6	23.4	8.7	19.4	13.7	7.2	11.8	11.2	13.1
Merchandise trade (% of GDP)	Ghana	56.50	37.32	39.23	57.18	53.70	32.92	35.71	56.21	93.40	69.91
	Uganda	52.09	41.51	36.05	19.67	51.26	20.29	10.22	26.34	33.68	30.17
Gross fixed capital formation (% of GDP)	Ghana	..	..	12.01	11.62	6.10	9.53	14.39	21.13	24.00	29.00
	Uganda	..	12.35	13.34	7.25	..	8.73	12.70	16.37	19.64	20.95
Gross domestic savings (% of GDP)	Ghana	17.14	8.29	12.78	13.66	4.94	6.64	5.47	11.59	5.34	3.43
	Uganda	16.49	12.46	17.24	5.79	-0.43	7.46	0.58	3.37	8.10	7.15

Source: World Development Indicators, 2007 (World Bank, 2007a).

Table A.2 Ghanaian and Ugandan economic statistics from national sources

		1975	1976	1977	1978	1979	1980	1981	1982	1983	1984
GNP changes year on year (%)	Ghana	-12.7	-3.7	+2.8	+8.6	-2.6	-0.7	-2.8	-7.0	-4.4	+8.2
GDP changes year on year (%)	Uganda	-2.0	+0.8	+1.6	-4.5	-11.9	-3.4	+3.9	+6.9	+5.0	-5.1
Consumer Price Index (1977=100) change year on year (%)	Ghana	29.7	56.3	116.3	..	..	50.1	116.5	..	..	139.6
GDP deflator (1991=100) change year on year (%)	Uganda	..	..	..	..	..	..	..	138.2	136.2	164.5

Sources: Ghana: Quarterly Digest of Statistics (various issues); Uganda: Background to the Budget (various years).

6 The global economy and developing countries

6.1 Introduction

Few issues have excited more attention in recent years, amongst academics, politicians and the media, than the question of globalisation. For some, the process of globalisation is inevitable and irreversible, something that national economies are forced to adapt to in order to survive and prosper in an increasingly integrated and competitive global economy. For others, globalisation is seen as a threat to the survival of the nation state and a constraint on the possibilities for economic development. There are no universally agreed definitions as to what globalisation actually is, there is little agreement on how economic statistics can be used to 'measure' globalisation and there is no consensus as to whether the current globalisation process is radically different from previous historical episodes. Globalisation is not the inevitable result of 'natural' or technical developments in the global economy. Globalisation is in fact a policy-induced process – national governments, either unilaterally or collectively, and sometimes perhaps acting under duress, agree to create the institutions (the World Trade Organization, for example) or take the decisions (financial sector liberalisation) that together make up our notions of globalisation. Technological changes – developments in ICT, more rapid air transportation (for both people and freight) – mean that the importance of geographical distance may well be decreasing, the earth may be becoming 'flatter' and the world 'shrinking'. But the nation state is not in danger of disappearing or becoming irrelevant. Indeed, in a more interdependent and uncertain world, 'the nation state may become more salient as a means of protection against global forces beyond supranational control' (Hirst and Thompson, 2003: 34).

In this chapter, we consider first the various dimensions of globalisation, outline the inequalities in the contemporary global economy and discuss the problems involved in determining whether global economic growth is convergent or divergent. We then briefly discuss the institutions of globalisation, global instability and the impact of globalisation on poverty alleviation.

6.2 What is globalisation?

Globalisation is a multi-dimensional process of transformation which has many different meanings and interpretations, both economic and non-economic. It generally refers both to the increasing flows of capital, goods, and resources and knowledge across national boundaries and to the emergence of a complementary set of organisational structures to manage the expanding network of international economic activity and transactions. A stricter definition of globalisation would be the emergence of a global economy where firms and financial institutions operate transnationally – that is, beyond the confines of national boundaries. Differing interpretations of globalisation carry varying emphases.

- In one interpretation, there is the development of global capital markets and large and volatile movements of short-term capital, the result of the liberalisation of balance-of-payments capital account transactions (as in the build-up to the 1997 Asian financial crisis).
- In another interpretation, there is the domination of the global economy by huge transnational corporations (TNCs) in key economic sectors – automobiles, media, entertainment and leisure, pharmaceuticals, financial services, telecommunications and information and communication technology (ICT).
- Others (O'Neill, 1997) point to the increased interdependence between economies through international trade and direct foreign investment (DFI) (so-called 'shallow integration').
- Then there is the globalisation of consumption through the development of global brands and trademarks – soft drinks, cigarettes, clothing and footwear, music, the media.

- Others (the OECD, for example) stress the emergence of cross-border value-adding activities and technology transfer within TNCs and within networks established by TNCs which reinforce linkages between national economies (the development of global value chains in key sectors, for example, clothing). This is so-called 'deep integration' (O'Neill, 1997) in which technological change in general, and the development of ICT in particular, plays key roles. Note that this interpretation of globalisation sees it as a microeconomic phenomenon, rather than as macroeconomic change.
- Broader conceptualisations of globalisation would encompass issues relating to the environment (global warming), security, illegal narcotic drugs, the spread of diseases (bird and swine flu, for example) and the spread of liberal democracy.

Writers such as Hirst and Thompson (1996) are critical of the more extreme versions of globalisation, such as those predicting the end of the nation state. They argue that: a) a highly internationalised economy is not unprecedented, because the late nineteenth and early twentieth centuries were also a period of rapid global change; b) genuinely transnational corporations with no national base are rare; and c) there has been no massive shift of investment and employment to developing countries (that is, there is no 'new' international division of labour).

It is also necessary to point to the fact that, on any definition, the process of globalisation has not led to a fully functioning 'global economy'. There is no global labour market (although labour migration and the flow of remittances to the country of origin is becoming increasingly important for many countries – for example, Bangladesh and the Philippines), except perhaps in certain occupations or activities, such as medicine, software engineering, academia and football. There is as yet no global taxation system aimed at improving the efficiency and fairness with which global resources are allocated and generating revenue for the provision of global public goods. There is also no global system of regulation with respect to market structures and the behaviour (conduct) of global enterprises, although 'ethical behaviour' by TNCs is seen as an increasingly important goal. The Geneva-based United Nations Conference on Trade and Development (UNCTAD, 1997) has argued for the use of the term 'global economic interdependence' rather than globalisation, to indicate that the latter is not complete and that nation states retain distinct economic identities with decision-making powers. UNCTAD argues that global economic forces can clearly bring potential benefits to low-income economies. The main issue is how to manage the interaction of domestic and international economic forces so that global economic interdependence leads to faster economic growth and rising living standards in those countries.

The International Monetary Fund (IMF, 1997) has argued that countries that aligned themselves with the forces of globalisation and implemented the necessary economic reforms were likely to converge with the advanced economies (the Republic of Korea, Singapore, Taiwan and Israel are quoted as examples). Countries which do not adopt policies consistent with globalisation were likely to face declining shares of world trade and private capital flows and would fall behind, in relative terms, the advanced countries. What the IMF in effect is saying is that if countries do not benefit from globalisation, it is their responsibility.

The UNDP (1999: 43–4) argues that:

> Globalization expands the opportunities for unprecedented human advance for some but shrinks those opportunities for others and erodes human security. It is integrating economy, culture and governance but fragmenting societies. Driven by commercial market forces, globalization in this era seeks to promote economic efficiency, generate growth and yield profits. But it misses out on the goals of equity, poverty eradication and enhanced human security.

Box 6.1 demonstrates that the empirical evidence is ambiguous on whether globalisation is leading to economic convergence; that is, reductions in differences between levels of per capita income or productivity in rich and poor countries. Some countries have 'caught up' over a long time period (Japan, the countries of Southern Europe, some Arab states, Singapore, Hong Kong, the Republic of Korea and a number of other East and South-East Asian economies). However, divergence in the form of increasing inequalities between rich and poor countries, appears to be more likely for most economies. Whether this is effectively inherent in the globalisation process itself (i.e. it excludes poor countries as the UNCTAD and UNDP seem to imply) or that it is poor policy in those countries that is to blame (the IMF's position) remains an open question. In this context, the

International Labour Office (ILO) issued a Report (ILO, 2004) highlighting the dangers of the increasing gaps between the rich and the poor of the world.

Box 6.1

Measuring world income distribution

World income distribution is a combination of a) the internal income distributions of all countries and b) the distribution of average incomes between countries. Most of the inequality in world income distribution reflects inequalities between country averages rather than inequality within countries. World income inequality is very high, with an estimated Gini-coefficient of 0.66 using incomes adjusted for differences in consumers' purchasing power (purchasing power parity estimates – PPP) and 0.80 if current dollar incomes are used. These are controversial figures and although there is disagreement on the matter, new empirical evidence suggests that world income distribution is rapidly becoming more unequal. The increase in world inequality between 1988 and 1993 (from 0.628 to 0.660) occurred as a result of both between-country and within-country increases in inequality. The slow growth of rural per capita incomes in populous Asian countries (China, India and Bangladesh) compared to the income growth of several large and rich OECD countries, plus the fast growth of urban China compared to rural India were the main reasons why the world Gini-coefficient increased. What happens to world income distribution depends to a large extent on what happens to the relative position of China and India at one end of the spectrum and the USA, Japan, France, Germany and the UK at the other end.

Estimates of world income distribution depend heavily on:

- the measure of inequality used;
- the weight given to individuals or countries;
- the use of currency market exchange rates or PPP exchange rates to compare incomes in different countries.

If countries are treated equally (i.e. not weighted by population) and if average income is measured in PPP prices, world income distribution has become more unequal over the past few decades whatever measure of inequality is used.

If countries are weighted by their population, world PPP income distribution has shown little change. When incomes are expressed in a common currency using market exchange rates, evidence shows that world income distribution has become much more unequal over the past few decades and that inequality accelerated over the 1980s and 1990s, irrespective of whether countries are

continued

treated equally or weighted by their population. Wade (2001b: 8) concludes that 'when countries' incomes are compared using market exchange rates, globalisation has not worked, in the sense that rapidly increasing flows of trade and investment have not yielded the expected neoclassical convergence, have not benefited the poorer participants by nearly as much as the richer'.

These are disputed conclusions. There are severe problems with estimating and using PPP prices and there are disagreements over the estimates for Chinese Gini-coefficients. But there is some agreement that world income distribution matters (Milanovic, 2007: 154). Globalisation has increased the awareness of inter-country differentials in income and wealth and leads to increased pressures for migration of labour, both legal and illegal. High and possibly increasing levels of world income inequality are also inconsistent with notions of international economic justice.

6.3 Global economic inequality

The process of globalisation impacts on income distribution both within a country and between countries. Within a country, for example, the sectors and factors of production which are incorporated into the global economy through direct foreign investment may benefit economically. This benefit applies even though labour working in such sectors may be paid comparatively low wages, although these wages may be higher than would be received in the agricultural sector. Income distribution between countries will in part be determined by relative rates of economic growth, which in turn are in part determined by levels of investment, flows of technology, educational levels, and appropriate government policies, including the promotion of innovation and entrepreneurship.

It is easy to exaggerate the changes that have occurred globally. As Dicken (2007: 68) notes:

> there have been big changes in the contours of the global economic map. But the fact remains that the actual extent of global shifts in economic activity is extremely uneven. Only a small number of developing countries have experienced substantial economic growth; a good many are in deep financial difficulty whilst others are at, or even beyond, the margins of survival . . . Although we can indeed think in terms of a new international division of labour, its extent is far more limited than is sometimes claimed.

Even though there have been significant changes in the global economy over the past two decades, the older 'core' economies still dominate international trade and investment flows, and the global economy is still highly unequal. Patterns of development are highly uneven in terms of the global distribution of income, of manufacturing production services and of international trade (Dicken, 2007: Chapter 2 – this section draws heavily on Dicken's work). The concentration of economic activity in the 'core' developed market (capitalist) economies has been described as the 'global triad'. North America, Europe and East and South-East Asia together account for 86 per cent of world GDP and world merchandise exports, 75 per cent of inward direct foreign investment (DFI) inflows and 98 per cent of DFI outflows. Three countries alone – the USA, Japan and Germany – still account for over 50 per cent of global manufacturing value added (MVA). The USA, in addition to its 25 per cent share of global MVA, also accounts for 33 per cent of commercial services and 15 per cent of agricultural production.

The major change that has occurred in the global economy over the past forty years has been the emergence of the East and South-East Asian economies – Republic of Korea, Taiwan (sometimes referred to as Taiwan Province of China), Hong Kong (the Hong Kong Special Administrative Region since 1997), Singapore – and the 'second tier' (or newly industrialising economies – NIEs) usually comprising Thailand, Malaysia, Indonesia and sometimes the Philippines. The (re-)emergence of the People's Republic of China since the early 1980s is not only remarkable in itself but also has huge implications for the global economy in general and the prospects for the development of the low-income economies (not just in Asia) in particular.

Some economists (for example, Kaplinsky and Messner, 2008) refer to these Asian economies (along with India and Vietnam) as the 'Asian Drivers', economies that will have a significant impact through their growing competitiveness, their size and their rapid economic growth. Other economists follow the terminology introduced by Goldman Sachs and refer to the BRICs – Brazil, Russia, India and China. The focus on China is usually in its role as an exporter of largely low-cost manufactured goods, especially textiles and clothing items, and their impact (the competitiveness impact) on other low-income exporters. With the ending of the Multi Fibre Agreement (MFA) in 2005, textiles and clothing exporters in low-income economies lost their privileged access, through quotas, in the developed market economies and China was the major beneficiary of this change. But there are also complementary effects – China already out-sources many activities to other Asian economies, and is an increasingly important market for foodstuffs, raw materials and fuels imported from a number of sub-Saharan and Latin American economies. The impact of China on individual economies therefore depends on the balance between these competitive and complementary impacts.

The nature of the triad has thus changed with the addition of the new East and South-East Asian 'member' economies. The important point remains, however, that a large proportion of trade and DFI flows take place within these regions (intra-regional) and some economists see greater regionalisation (a market-driven form of regional integration within RTAs – regional trading areas) as a process that will perhaps offset or alleviate some of the more negative effects of globalisation.

Globalisation and regionalisation may thus essentially consist of greater economic interdependence both between and within the three (augmented) pillars of the triad, accelerated by further economic liberalisation, deregulation and the privatisation of economic activities. But this is a process that may well continue to exclude large parts of the developing world, especially the majority of sub-Saharan economies and large parts of Latin America, what Dicken (2007: 47) calls the 'persistent peripheries'. It is not surprising that those economies that are unable, for whatever reason, to exploit the opportunities that it does provide, and see only its destructive impact (on previously import substituting industries as a result of trade liberalisation, for example), have a more negative and pessimistic view of their future in an increasingly integrated global economy.

Global income distribution

We have argued above that the process of global economic growth is more likely to be diverging than converging. Given that we are fairly confident that income distributions within countries are becoming more unequal, it is reasonable to conclude that inequalities in the global distribution of income are also increasing over time (see Box 6.1).

The UNDP (1999) noted that world inequalities have been rising for nearly two centuries. The distance between the richest and poorest countries was about 3 to 1 in 1973 and 72 to 1 in 1992. Wade (2001a) also argues that world income distribution became more unequal over the last two decades of the twentith century. The World Development Report 2006 (World Bank, 2005: Chapter 3) notes the conflicting empirical evidence on global income inequality and the report as a whole focuses attention on equality of opportunity, especially with respect to health and education.

It is in fact very difficult to determine what is happening to global inequality. Sutcliffe (2007) shows that the conclusions that we reach depend upon a) what we are measuring, b) how we are measuring it, c) where the data come from and d) the measure of inequality used. Overall global inequality depends on many inequalities relating to specific issues and which cannot be summarised in a single measure.

6.4 The institutions of globalisation

The major institutions of the globalised (or of the increasingly economically interdependent global economy) are the IMF, the World Bank (WB) and the World Trade Organisation (WTO). An attempt was made to establish a Multilateral Agreement on Investment (MAI) under the auspices of the Organisation for Economic Co-operation and Development (OECD) in Paris in 1998 (Fitzgerald et al., 1998). That attempt failed, and responsibility for future efforts at creating a new MAI has been transferred to the WTO in Geneva.

Both the IMF and the WB were established at the International Monetary and Financial Conference of the United and Associated Nations at Bretton Woods (USA) in 1944. The IMF was to be mainly concerned with the provision of an adequate supply of international finance so that countries experiencing short-term balance-of-payments problems would not have to resort to protectionist measures, thus maintaining an effective mechanism for international adjustment and stability. Since the early 1980s, the IMF has devoted most of its attention and resources to low-income economies which have balance of payments problems, through stabilisation programmes negotiated with those countries. Special concessionary facilities were created for poor countries, renamed the Poverty Reduction and Growth Facility (PRGF) in 1999.

The World Bank (properly two legally and financially distinct entities – the International Bank for Reconstruction and Development (IBRD) and the International Development Association (IDA)) was concerned initially with reconstruction of the war-torn economies of Europe, but from 1949 onwards its priorities shifted towards the promotion of economic development in what became known as the 'Third World'. The IDA was established in 1960 as the soft-loan arm of the IBRD. 'Soft loans' are those with low rates of interest, moratoria for capital repayments and interest payments, and long repayment periods. Since the 1980s, the WB has increasingly been concerned with programme loans to poor countries (previously its activities had largely been based on projects) under the general headings of structural adjustment, or sectoral adjustment, programmes. Some further discussion of what is known as 'aid architecture' will be found in Chapter 8.

In reality, the two sets of programmes appear to be similar. Both the IMF and the WB attach conditions to their loans to ensure that recipients meet certain agreed targets, pursue specified policies or implement agreed institutional changes. There is often 'cross-conditionality' between the two institutions (that is to say that the WB will not agree to a loan unless there is already an IMF stabilisation programme in place).

A typical stabilisation programme focuses on the management of aggregate demand and will consist of:

- monetary and fiscal contraction to reduce the public sector budget deficit in order to reduce aggregate demand (deflation);
- devaluation of the exchange rate (now more usually a precondition for the agreement of the signing of a Letter of Intent with the IMF);
- liberalisation of the economy via the elimination of controls and regulations, privatisation, trade liberalisation (removal of quotas and rationalisation and reduction of tariffs);
- wage restraint, removal of subsidies and reduction of transfer payments (but often accompanied by establishment of a 'social safety net').

A typical structural adjustment programme (which attempts to shift the aggregate supply curve of the economy) may well include all or some of the above measures, but will also include:

- measures to strengthen capacity to manage the public investment programme;

- revised agricultural pricing policies;
- revised industrial incentives;
- reform of the budget and tax system;
- improvement of export incentives.

In 1999, the IMF established its Poverty Reduction and Growth Facility (PRGF) to provide a low-interest lending facility for low-income countries to make the objectives of poverty reduction and growth more central to lending operations. The seventy-eight low-income countries that are eligible to borrow from this facility now do so within the framework of a Poverty Reduction Strategy Paper (PRSP – resulting in a Poverty Reduction Strategy) agreed with the World Bank but 'owned' by the borrowing country (IMF, 2009; World Bank, 2002). Many countries have developed PRSPs in collaboration with the IMF and the World Bank. PRSPs consist of a broad framework projecting macroeconomic performance into the future, a summary of anticipated government finances for a period of at least three years, and a brief statement of major sectoral policies and developments. Release of debt relief funds under the Heavily Indebted Poor Countries (HIPC) scheme was conditional upon countries having an agreed PRSP. In many respects, the preparation and publication of PRSPs has represented the reintroduction of a form of economic development planning.

The economic impact of IMF and the WB programmes of economic reform has been the subject of heavy criticism, not all of it justified. These institutions will continue to evolve and adapt over time in response to the changes in the global economy and in order to try to resolve perceived anomalies in this process. In early 2009, as the global financial crisis unfolds, calls are already being made for a substantial expansion of IMF resources and, more radically, a change in its constitution that will bring perhaps countries such as India and China more fully into the decision-making structure of the organisation.

6.5 Global instability

The small economic size and limited economic influence of most poor countries means that they are highly vulnerable to developments within the global economy which are beyond their control but which may have an adverse impact on their income, employment and prospects for economic growth.

The past four decades have been characterised by instability in the global economy:

- The early to mid-1970s witnessed the breakdown of the post-1945 system of fixed exchange rates (the Bretton Woods system), followed by two oil price shocks (1974/75 and 1979/80) and by a combination of inflation and recession (so-called stagflation).
- The 1980s were characterised by slow economic growth, continued economic instability and growing disparities in economic performance between countries. The decade ended with the 'revolutions' in eastern and central Europe in 1989, and in 1990 there was the dissolution of the Council for Mutual Economic Assistance (the CMEA or COMECON), which was the 'common market' between the Soviet Union and its eastern and central European allies.
- The 1980s was also a decade of crisis for the developing countries as a whole, and the period is often referred to as the 'lost decade'. It was subjected to a series of major external shocks (see Box 6.2) in the early 1980s, and the continuation of adverse factors throughout the period made economic recovery difficult for many countries.
- The global economy is again subject to major economic instability resulting from the US and UK banking crises of 2008 and 2009 (see Box 6.3).

Excluding the transitional economies, however, there was an improvement in the growth performance of poor countries in the 1990s, as noted above. This was largely the result of the rapid growth of the East and South-East Asian economies.

The Asian financial crisis of 1997 and other crises in Mexico (1995, the so-called Tequila Crisis), Russia (August 1998), Brazil (January 1999) and Argentina (2002) introduced new elements of instability in the global economy, largely the result of many countries liberalising their balance-of-payments capital accounts and thus becoming vulnerable to massive short-run inflows and outflows of short-term capital (so-called 'hot money').

The Asian economies in general have recovered from the crisis, and the fear of contagion from the Russian, Brazilian and Argentinian crises appears in retrospect to have been exaggerated. 'Contagion'

Box 6.2

What are external shocks?

The external shocks that most affected the developing countries were largely caused by the slowdown in global economic activity, especially in the mid- and late 1970s (the global oil price shocks) and the early 1980s (tight monetary policy and higher interest rates in the developed market economies), which affected economic and industrial development through a number of experiences:

- A reduction in demand for the developing countries' commodity and mineral exports.
- A fall in commodity prices and a reduction in the net barter terms of trade (explained in Chapter 7).
- A significant rise in global real interest rates,[1] leading to an increase in the real burden of interest and debt repayments of poor debtor countries (the beginnings of the global debt crisis in 1982 with the near default of Mexico).
- A fall in the quantity of official development assistance (aid) and other capital flows to developing countries, resulting in negative net transfer of resources in 1984 and subsequent years for Latin America and Africa.
- Increased vulnerability to rapid and significant changes in short-term capital movements in the late 1990s.

External shocks in the early twenty-first century are manifesting themselves through higher oil, raw materials and food prices, slower global economic growth, instability in both national and global financial markets and increasing evidence of the deleterious effects of global climate change.

---

refers to the speed and ease with which economic and financial problems spread from one country to another. In an increasingly globalised world, 'contagion' has become a more serious problem with financial imbalances and inflation being transmitted globally considerably more quickly than in the past. But underlying problems persist. There remains potential instability inherent in the dependence of so many developing countries on volatile foreign capital inflows (UNCTAD, 1999).

6.6 The Asian financial crisis

The Asian financial crisis of 1997 was one of the most serious financial crises since the end of the Second World War. It was largely unexpected, hit a number of rapidly growing and highly successful economies, and promoted the largest financial bailout in history, stretching IMF resources to their limit.

The crisis began in Thailand in July 1997 with the devaluation of the Thai baht, following downwards pressure on the baht and the collapse of a number of key companies in the Thai financial sector. The contagion quickly spread to a number of other economies – the Philippines, Malaysia and Indonesia and finally the Republic of Korea – and exchange rates were floated and allowed to depreciate. Other Asian economies – India, Pakistan, China and Taiwan – were either less affected by the crisis or not affected at all. The IMF intervened early in the crisis and agreed rescue packages with Thailand, Indonesia and Korea. Malaysia did not agree to a programme with the IMF; and the Philippines, less affected than the other countries, was already in an IMF programme which was subsequently extended. Singapore and Hong Kong Special Administrative Region were hit by the crisis in 1998.

There are two basic interpretations of the causes of the crisis:

- The majority view (including that of the IMF) focused on structural problems and fundamental weaknesses in the Asian economies – large and persistent trade and current account deficits, loss of external competitiveness in key export markets, excessive government intervention in these economies, 'crony capitalism' (unhealthy, close relationships between enterprises and governments), and problems of moral hazard (in this case where the incentives surrounding actions are distorted by the existence of explicit or implicit guarantees against loss – the term 'moral hazard' is explained in Chapter 8).
- The second interpretation argues that although there were weaknesses and underlying problems in the Asian economies, they were essentially sound but under-regulated, especially with respect to their financial sectors, and that it was financial panic that was the basic element in the Asian crisis. Panic among the international investment community, policy mistakes at the onset of the crisis by Asian governments, and poorly designed international rescue

programmes turned short-term capital inflows into a fully fledged financial panic and deepened the crisis.

The countries most affected by the crisis (Thailand, Malaysia, Indonesia and Korea) had liberalised capital account transactions and deregulated their financial sectors without putting in place the necessary framework of prudent regulation. Net private capital flows into the five most affected countries had risen from US$37.9 billion in 1994 to US$97.1 billion in 1996. These inflows suddenly reversed in the second half of 1997, turning to an outflow of US$11.9 billion, a turnaround of approximately US$109 billion (about 10 per cent of the pre-crisis national income of these economies). Economies that had not practised capital account liberalisation of transactions (China) or only partially liberalised them (Taiwan) were less affected by the crisis.

The consequences of the crisis included:

- a negative impact of the external trade of these economies (with knock-on effects for their major trading partners, that is, a slowdown in international trade);
- falls in output, increases in bankruptcies and rises in unemployment;
- sharp rises in prices, especially those for basic necessities (foodstuffs and medicines);
- cuts in public expenditure as part of IMF stabilisation programmes;
- the erosion of the social fabric – political and social unrest in Indonesia, for example, increased poverty and inequality, with rises in crime.

The role of the IMF has come under increased scrutiny as a result of the crisis. Critics have argued that it misinterpreted the causes of the crisis, insisted on a policy package to rescue the economies which was too deflationary, and in effect (some would say with the backing of the US Treasury) forced these countries, especially Korea, to deregulate and liberalise their economies, especially the financial sector, in order to encourage greater openness and foreign investment (for example, Stiglitz, 2002: Chapter 4). The heavy indebtedness of the Korean chaebol (large, diversified business groups such as Samsung and Daewoo) has already led to significant sales of assets to foreign investors and massive restructuring. It is unlikely that the Korean state will be able, or wish, to play as much of a significant developmental role in the future as it has in the past. There has been a massive

Box 6.3

The impact of the global financial crisis of 2009 on less developed countries

As the global financial crisis unfolds, increasing attention is being focused on its likely impact on developing countries. The World Bank (WB) (2009) estimates that 53 million people could be trapped in poverty as economic growth slows around the world, in addition to the 130–155 million people pushed into poverty in 2008 because of rising food and fuel prices. It is estimated that 94 out of 116 developing countries have experienced a slowdown in economic growth. The World Food Programme (WFP) of the United Nations argues that hunger is 'back on the march' with 115 million people added to the ranks of the hungry over the past two years, driven by macro factors, the food crisis, the financial crisis and climate change. As of February 2009, global grain prices were 13 per cent higher than one year ago and 83 per cent higher than in 2005.

The IMF (2009) expects the global crisis to have a major impact on low-income countries, especially in sub-Saharan Africa. The IMF uses a concept of vulnerability that is based on an assessment of a country's overall level of exposure (defined by a situation where the initial level of poverty was a problem before the crisis and where an adverse impact on economic growth is expected) to argue that the crisis is exposing households in virtually all developing countries to increased risks of poverty and hardship. Almost 96 per cent of countries are either highly or moderately exposed according to WB calculations.

The United Nations Educational, Social and Cultural Organization (UNESCO) is concerned with the impact of the global crisis on human development, including an increase in infant mortality, child malnutrition and the failure to meet the educational targets enshrined in the Millennium Development Goals (MDGs).

Overall, there is a high degree of unanimity about the impact of the crisis. Those countries most integrated into the global economy through trade, DFI and remittances are likely to suffer most. Demand for exports will fall, remittances will fall, and DFI flows and aid are under threat. The budgetary position of many governments will also worsen (slower economic activity, lower commodity prices, potential falls in aid) at a time when many countries will need to increase spending to protect the poor and service higher debts.

The IMF urges countries to maintain macroeconomic stability, with increased financial support from the multilateral institutions; the WB calls for both increased multilateral and bilateral donor commitments, including a Global Food Response Program and the WFP calls for food security to be kept high on the agenda and highlights a number of countries at risk, including Zimbabwe, Somalia, Pakistan, Afghanistan, Bangladesh and Haiti.

increase in inflows of direct foreign investment into Korea and Thailand in particular as foreign companies buy up relatively cheap assets and/or invest in sectors of the economy from which they were previously denied entry. Rodrik (2009) argues for strong national regulatory mechanisms, with a 'thin veneer of international co-operation'.

The IMF has defended its role and has pointed to the stabilisation of exchange rates and the rapidity of the recovery as evidence of success. It is also the case, of course, that we do not know what would have happened if there had been no IMF or IMF intervention. We have no counterfactual, that is, what might have been the overall situation if the IMF had not intervened in the manner that it did. We cannot rerun history to answer such a question. The concept of the 'counterfactual' is explained in more detail in Chapter 8.

6.7 Globalisation and poverty alleviation

Globalisation offers new opportunities to countries through closer integration into the global economy, but the actual distribution of benefits may be very unequal and it cannot be assumed automatically that the poor will necessarily benefit from it. In addition, the policy reforms that constitute globalisation severely reduce the autonomy of national governments and make the emergence of a developmental state more difficult to achieve. Policy choices, for example with respect to trade, fiscal and monetary policy, which both directly and indirectly impact on the achievement of poverty alleviation targets, are often constrained by the demands of global institutions. There is no agreement in the literature on the overall impact of globalisation on poverty and the focus of attention remains on the variety of channels and transmission mechanisms through which the process of globalisation affects different aspects and dimensions of poverty (Thorbecke and Nissanke, 2006: 1333).

The main channels identified are drawn from what we referred to above as the differing interpretations of globalisation, namely openness and trade, financial liberalisation, technology transfer, factor mobility, vulnerability and flows of information and institutional change.

The debate on trade liberalisation ('openness'), economic growth and poverty alleviation remains unresolved. Neo-classical trade theory hypothesises that with free trade, scarce resources will be re-allocated in a manner consistent with a country's comparative advantage. In a labour surplus–capital scarce economy, other things being equal, this implies the creation of more employment opportunities in relatively labour-intensive sectors or activities.[2] Some empirical evidence seems to suggest the opposite outcome, however. In some Latin American economies, trade liberalisation has led to growth in demand for skilled labour, increasing inequality between wage earners. Technologies transferred and diffused through globalisation may be relatively capital intensive and thus may have a relatively limited impact on net employment creation. Financial sector liberalisation may increase an economy's vulnerability to external shocks (see above) with a negative impact on poverty.

Ultimately it is national policies, not globalisation as such, that determine the rate and characteristics of economic growth; that is, whether it is pro-poor or not. The eminent economist Dani Rodrik (2007: 2) makes the point well – 'national policy choices are the ultimate determinant of economic growth. At the same time, successful countries are those that have leveraged the forces of globalisation to their benefit'. Other economists (for example, Cornia, 2004) argue that increased income inequality within countries, reducing the effectiveness of poverty alleviation measures, is attributable to the nature of technological changes and policy reform measures (stabilisation and structural adjustment programmes), along with privatisation, changes in labour institutions and the unwillingness or inability of states to pursue redistributive objectives.

6.8 Globalisation, underdevelopment and 'dependency'

Globalisation raises strong passions both for and against. Those who emphasise the negative impact of globalisation on LDCs can, if they are aware of its existence, refer back to a body of earlier, radical literature which analysed the position of LDCs within the global economy and, in particular, focused on their historical experience of colonialism. In this penultimate section of the chapter, we take the opportunity to set these contemporary concerns within the context of this earlier literature. The terms 'underdevelopment' and 'dependency theory' tended to be used interchangeably to refer to a body of work, both theoretical and empirical, largely, but not exclusively, Latin American in origin which argued that the process of development in the capitalist centre of the global economy led to a process of underdevelopment in the periphery (developing countries). 'The development of underdevelopment' was the slogan that best expressed this interpretation of global history and the impact of capitalist economic growth: a slogan coined by a Chicago University-trained North American economist, Andre Gunder Frank (1969).

Latin American structuralist and dependency writers (including Caribbean writers) tended to focus on the behaviour of the terms of trade (see Chapter 8) and the need for the development of a manufacturing sector in the peripheral economies. However, with the apparent failure of import substituting industrialisation (ISI) (see Chapter 3), a theory of dependent development emerged in which the Brazilian sociologist Teotonio Dos Santos (1973) argued that dependency entailed the development of a 'periphery' which was conditioned by the capitalist core economies. Cardoso and Faletto (1979) – the sociologist F.H. Cardoso later became president of Brazil – outlined a theory of 'situations' of dependent development in which the dependent economies lacked a machine-making capacity (they had no capital goods sector and were thus technologically dependent and dependent upon capital goods imports) and also lacked the ability to internalise the system of capital accumulation. But dependent development was a possibility; that is, dependent economies were not necessarily becoming poorer or 'underdeveloping'.

Many of the underdevelopment and dependency theorists were described as 'Neo Marxists' to distinguish them from 'classical' Marxism (Marx, 1853). In his writings on India, Marx had distinguished between the destructive (plunder and annihilation of Asiatic societies) and the constructive (political unification, the establishment of private property, technology transfer – the development of the railway system in India) impact of colonialism, with the latter eventually leading to the foundations of a capitalist economy in the colonies. A leading North American Marxist economist (Baran, 1957) argued that the capitalist penetration of the periphery developed some of the prerequisites for capitalist development but blocked others. Although many of those engaged in current debates over the impact of globalisation may not realise it, the basic question remains: can the capitalist mode of production (the free market economy in other words) actually establish itself in developing countries? The answer may be affirmative in large parts of East and South-East Asia but negative so far in most of sub-Saharan Africa.

6.9 Summary

- The small economic size and limited resources of the majority of poor countries means that they exert little influence on the direction that changes in the global economy are taking, and it also makes them extremely vulnerable to adverse changes in global economic conditions and exogenous shocks of all kinds.
- Some economies have been able to adjust to changed circumstances and have sustained long periods of rapid economic growth, while others have been unable to adjust and face economic stagnation and social and political crisis.
- Economists disagree as to what factors determine a country's ability to adjust. Orthodox economists emphasise the importance of free trade and the market in reallocating resources and making an economy open to competitive pressures, but others point to the role of the 'developmental state' in the East Asian context and emphasise the importance of selective intervention by governments to ensure the achievement of development objectives.
- Development economists quite properly continue to focus attention on issues of global poverty, inequality and instability, but it would be misleading to ignore the profound changes that are occurring in the global economy and their impact on, and implications for, economic development.
- There is great diversity within the 'developing world' with respect to history, political evolution, institutions, and growth and development performance.
- The forces of globalisation, and of greater economic interdependence, are creating both new opportunities and new problems for less developed countries. New policies are needed to take advantage of those opportunities and new institutions, both national and global, are needed to deal with the new problems which will arise.
- At the time of writing (early 2009) calls are being made to expand and reform the IMF and other international financial institutions, along with calls for the creation of new global regulatory structures for the financial and banking sectors. This is consistent with an interpretation of globalisation that sees it not as a supranational process beyond the control of nation states but one determined by nation states themselves.

Questions for discussion

1 'Globalisation is a process that is both inevitable and irreversible'. Discuss.

2 Distinguish between shallow and deep integration in the globalisation process. Which is the most important in your opinion?

3 What are the major problems economists face in trying to determine changes in world income distribution?

4 Why are less developed countries vulnerable to external shocks? What is the likely impact of the 2008–9 global financial crisis on economic development?

5 The IMF (2009) has called for the international community 'to act urgently and generously to avoid the potentially devastating effects of this crisis [2009] on the most vulnerable economies'. What effective steps can the international community realistically take?

Suggested further reading

Dicken, P. 2007. *Global Shift: Mapping the Changing Contours of the World Economy* (5th edn). London: Sage.

Hirst, P. and Thompson, G. 1996. *Globalization in Question*. Cambridge: Polity Press.

ILO 2004. *A Fair Globalization: Creating Equal Opportunities for All.* Geneva: International Labour Office for the World Commission on the Social Dimensions of Globalization.

Krugman, P. 2008. *The Return of Depression Economics and the Crisis of 2008*. Harmondsworth: Penguin Books.

Nixson, F. and Walters, B. 1999. The Asian Crisis: Causes and Consequences. *Manchester School*. 67 (5): 496–523.

O'Neill, H. 1997. Globalisation, Competitiveness and Human Security: Challenges for Development Policy and Institutional Change. *European Journal of Development Research*. 9 (1): 7–37.

Stiglitz, J. 2007. *Making Globalization Work*. Harmondsworth: Penguin Books.

Wade, R. 2001c. The Rising Inequality of World Income Distribution. *Finance and Development*, 38 (4) December. Accessible from: www.imf.org/external/pubs/ft/fandd/2001/12/wade.htm.

World Bank. 2006. *World Development Report 2006*. New York: Oxford University Press for the World Bank.

Useful websites

International Monetary Fund: www.imf.org
World Bank: www.worldbank.org
UNESCO: www.unesco.org
The Economist: www.economist.com/finance

Economic concepts used in this chapter

Globalisation/global economy
Transnational corporations (TNCs)
Information and communication technology (ICT)
Direct foreign investment (DFI)
Value chains
Remittances
Global public goods
Convergence/Divergence of economic growth
International Monetary Fund (IMF)
World Bank (WB)
World Trade Organization (WTO)
Global economic inequality/income distribution/poverty
Global Triad
Brazil, Russia, India and China (BRIC)
Multi-Fibre Agreement (MFA)
Aggregate demand/aggregate supply
Liberalisation/deregulation/stabilisation/structural adjustment
Poverty Reduction and Growth Facility (PRGF)
External shocks
Stagflation
Devaluation
Capital account liberalisation
World Food Programme (WFP)
Chaebol
Purchasing Power Parity (PPP) Prices
Gini-coefficient
Millennium Development Goals (MDGs)

Notes

1 Real interest rates allow for the effect of inflation on 'nominal' interest rates. In simple terms, if the nominal rate of interest is 10 per cent per annum and the rate of inflation is 5 per cent per annum, then the real interest rate is 5 per cent per annum (nominal interest rate minus the rate of inflation).

2 This process is explained in more detail in Chapter 7.

7 Developing countries and international trade

7.1 Introduction

This chapter focuses on some of the key economic issues associated with developing countries' role in international trade in products. It complements the chapters relating to structural change and to globalisation. Because it concentrates on trade in products, it does not attempt to cover the overall balance of payments (which includes trade in services as well as financial flows) and it does not aim to discuss international migration (which is important and which has significant economic and financial implications – including remittances).

Following this brief introduction the second section will set out some of the basic economic theory relating to international trade, and particularly the theory of comparative advantage. This lies at the heart of many of the controversies in the international development area. The third section will discuss trade liberalisation, which has assumed a central and controversial role, particularly relating to 'structural adjustment' and 'economic policy reform'. The fourth section will focus on controversies relating to the position of developing countries in international trade including the terms of trade between developing countries and developed market economies (which was introduced in Chapter 2) and on recent institutional developments.

7.2 Conventional international trade theory

The basis for many of the controversies surrounding trade between developing countries and developed market economies has been the concept of gains from trade due to specialisation according to comparative advantage. The theory for much of the conventional economic analysis originates in the writings of the classical political economists, and in Ricardo in particular. Summaries of the conventional neo-classical approach to international trade theory can be found in Todaro and Smith (2008: 599–605) and in Salvatore (2004 – particularly Chapter 2). The origins of the theory of comparative advantage are traced in a recent article by Keith Tribe (2006). For a clear exposition of the role of trade in economic development in the nineteenth and twentieth centuries, a paper by Kravis is a very good example of how 'classic' articles retain a freshness and relevance which transcends decades (1970). His analysis is essentially in the 'economic history' category of development economics, rather than in the drier quantitative category.

The argument for 'free trade' based on comparative advantage provided the intellectual underpinning for the abolition of the protectionist Corn Laws in the United Kingdom in the mid-nineteenth century, and Box 7.1 is intended to explain the basic economic propositions in this context. In the standard exposition, specialisation and trade according to comparative advantage would result in aggregate production being higher than in the case of no specialisation and trade – this being the 'gains from trade' – while the distribution of these gains between the trading partners would be based on the ruling international exchange rates. The basis of the theory of comparative advantage is that countries may specialise in the production and export of products for which they have a relative (rather than absolute) advantage – and import products for which they have a comparative disadvantage. The outcome from this specialisation and trade will be higher levels of international production, and the distribution of these gains to each individual country will depend upon the terms of trade (i.e. the rate of exchange of the products). It is significant that the phrase 'distribution of gains' is used here – echoing precisely the point made by Singer in his seminal contribution to the Prebisch–Singer thesis discussed in Chapter 2 of this book (Singer, 1950).

The Heckscher–Ohlin 'factor endowment' addition to the basic theory of comparative advantage states that countries would specialise in the

Box 7.1

The Corn Laws and comparative advantage

Diagrammatic presentation of the basic theoretical approach relating to the arguments surrounding the abolition of the Corn Laws in mid-nineteenth-century UK is shown below for both the 'without-trade' and the 'with-trade' situations. In the left-hand diagram (for the UK) the curved line represents the range of production possibilities given the availability of factors of production (e.g. labour and capital), the natural resource endowment (climate, soil conditions, etc.) and the level of technology. By comparison, the right-hand diagram refers to the United States. The hatched straight line (at a tangent to the curve) in the two diagrams shows a) the combination of wheat and cloth actually selected for production (at the point of tangency), and b) the rate of exchange between wheat and cloth in both production and markets (the relative costs/prices are shown by the slope of the line). It should be clear that the rate of exchange (the relative costs/prices) between wheat and cloth are significantly different in the two countries (i.e. the slopes of the two hatched lines are different).

If the UK increases production of cloth for export, it has to switch resources from production of wheat (importing wheat from the USA) to cloth because the production possibility curve is an 'efficiency frontier' (increasing one requires reducing the other in a 'trade-off' situation involving opportunity costs) – while if the USA increases production of wheat, it switches resources from the production of cloth (importing cloth from the UK).

As the UK moves to specialise in the production of cloth, the straight line (tangent) moves to the right (to the dotted tangent), with the slope becoming 'steeper' indicating a change in the rate of exchange (relative costs/prices) between wheat and cloth in both production and markets. As the USA moves to specialise in the production of wheat, its straight line (tangent) moves to

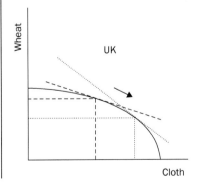

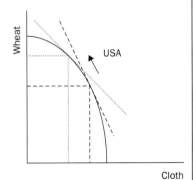

the left (to the dotted tangent), with the slope becoming 'shallower' indicating a matching change in relative costs/prices.

While this much-simplified exposition would probably not satisfy economic purists, it has the merit of demonstrating a number of important issues:

1 This approach is based on a process of abstraction involving simplifying assumptions and a clear cause–effect outcome which amounts to rigorous analysis.
2 The predictions of changes are made within a short-term 'comparative statics' approach – that is, other circumstances are unchanged (ceteris paribus).
3 In the longer term, the production possibility frontiers will move to the right as a result of a) investment, b) economies of scale, c) external economies, and d) technology change – this will also have the effect of changing the shape of the frontiers, significantly altering the outcome of the analysis – but in a manner which would be impossible to predict on a short-term basis.

The theoretical basis of the short-period economic analysis is based on neo-classical assumptions relating to market structures and to the nature of production (see Box 7.2). This can be summarised as the 2x2x2 framework of analysis (two countries, two factors of production and two products). For a critical elaboration of these theories, refer to Todaro and Smith (2008: 606–18), Thirlwall (2006: 514–22), Greenaway and Milner (1987) and Edwards (1985: 21–40). Kravis (1970: 857–8) has a fascinating discussion about the empirical impact of the abolition of the Corn Laws on production and trade in wheat for the UK, US and Europe.

production of products that depend mainly on factors of production which they possess in abundance. This means that countries would use factors of production for which their prices are low relative to international prices. The Samuelson–Stolper 'factor price equalisation' addition states that the long-term effect of using their abundant factors of production intensively would be to bid up prices, so that eventually countries which had benefited from relatively low-priced resources would lose this advantage – in the long-term there would be a tendency towards international uniformity of relative prices.

The policy implications of this approach would be for developed industrial countries to specialise in the production and export of more capital-intensive (and higher technology) products, and to import more labour-intensive (and lower technology) products produced and exported by developing countries. In the development economics literature, emphasis on trade policies which are based primarily on comparative advantage tend to be regarded as supportive of a *status quo* representing the interests of the developed market economies, while emphasis on trade policies which embody more 'structuralist' intellectual roots tend to be regarded as representing the interests of the developing countries. Both Todaro and Smith (2008: 606–18) and Thirlwall (2006: 542–4 and 555–7) set out the arguments for and against the free-trade approach to international trade policy. An important collection of essays covering many aspects of the international economic relations between developed industrial and developing countries, from a conventional perspective, is by Bhagwati (2002). For a structuralist critique of conventional theory, see Kay (1989 – particularly Chapter 2). H.-J. Chang has been a major protagonist on this issue in recent years, and his book *Kicking Away the Ladder* (2002) states his position very clearly.

Essential to the propositions of the theory of comparative advantage is the condition that in the 2x2x2 framework of analysis, which is explained in Box 7.1, both countries are producing both products, and that each country is relatively more efficient in the production of one of the products. However, many of the products which enter into trade between developing and developed countries are not produced in both groups of countries. This may be due to significant differences in natural resource endowments or in ecological (e.g. climatic) conditions. The former cannot be changed (such as mineral deposits) but the latter can be overcome (at a cost – e.g. greenhouses for horticultural production), which would justify the comparative advantage view. It is only in more recent decades, since many developing countries have developed the capacity to produce a wider range of manufactures on a competitive basis, that similar commodities have been produced very widely in both developing and developed countries. A well-known example of this in recent years has been the increased level of exports of horticultural products, including flowers, from developing countries to high-income industrialised countries – an increased source of income to some developing countries from 'non-traditional exports' which has significant disadvantages both for the exporting countries (diverting land and business competence from food production for example) and for the global economy (significantly increasing the amount of international air freight).

There are significant disagreements within the economics profession over the appropriate theoretical approach for the analysis of international trade, as well as over the interpretation of empirical data. Broadly speaking, the differing views fall fairly neatly into two broad schools of thought within each of which there are sub-groups.

First, is the neo-classical school (the 'conventional approach') which relies principally on marginal and short-term forms of analysis, tending to eschew more structuralist longer-term analysis, and to accept existing economic power relations and market asymmetries as given. Box 7.2 sets out the main assumptions associated with the characteristics of the neo-classical economic theory relating to the nature of perfect markets and of short-period production conditions. In this box, a comparison has been made with characteristics of imperfect markets and of long-period production theory which are not part of the neo-classical theoretical constructs. The neo-classical school tends to emphasise the limitations of government economic actions aimed to 'correct' for 'market failure', taking the view that such interventions often lead to 'government failure'. The issues of 'market failure' and 'government failure' are discussed in Chapter 8, along with the 'theory of second best', which provides an intellectual foundation for government intervention in markets. The neo-classical economic approach is often seen as tending towards 'neo-liberalism' and this issue will also be explored in more detail in Chapter 8.

Second is the structuralist school (which can be linked to 'heterodox economics'), emphasising the significance of diverse, asymmetric and 'imperfect' market structures in determining economic outcomes, and the need for institutional change in order to modify these outcomes in a purposive way. This school of thought tends to take a long-term view embracing the recognition of injustices within existing power relations but which sometimes uses marginal analysis in order to demonstrate the significance of economic structures in determining outcomes.

Structuralist economists are more prepared than neo-classical economists to accept that international trade works to the relative disadvantage of developing countries, and tend to take the view that determined international action is needed to reduce the asymmetries which underlie these disadvantages. Rather than relying on policy-induced fine-tuning (regarded as being 'neo-liberal') of international markets, the structuralist school would argue that strategic changes in

Box 7.2

International trade theory, 'perfect' markets and neo-classical short-period production theory

The elaborate structure of neo-classical international trade theory is based on assumptions which embrace a world with two countries, two factors of production (labour and capital), two products (one more labour intensive in production and the other more capital intensive), and with the same array of technologies available to producers in both countries (this can be described as the 2x2x2 framework). The other assumptions which lie behind the strict version of the theory are those relating to perfect competition, including free movement of products and factors of production within countries (but not between countries), perfect knowledge/perfect foresight, and freedom of entry to and exit from markets. The table which follows sets out the major assumptions or characteristics of a 'perfect market' together with a summary of the characteristics of an 'imperfect market' (or of the 'real world').

Characteristics of a perfect market	Characteristics of an imperfect market
1 Large numbers of sellers, none having the power to set market prices.	1 The market power of individual sellers varies in a continuum from single-firm industries (monopoly) to 'atomistic competition' in industries with many firms.
2 Large numbers of buyers, none having the power to set market prices.	2 The market power of individual buyers varies in a continuum from single-firm buyers (monopsony) to large numbers of individual consumers.
3 Free entry and exit to and from the market for producers and consumers.	
4 No transport or other 'transactions' costs – the markets have no 'spatial' characteristics.	3 Entry to and exit from industries is limited by factors including high investment costs, specialist technical knowledge (including copyright law) and by restrictive practices.
5 All sellers and buyers have perfect knowledge of all current market characteristics.	4 The information possessed by sellers and buyers varies considerably (i.e. asymmetric information).
6 All sellers and buyers have perfect foresight for all future market characteristics.	5 Sellers and buyers have imperfect knowledge about future market characteristics.

The second table, which follows, sets out the assumptions or characteristics of neo-classical short-term production theory, together with a comparative list of characteristics of long-term production theory.

Characteristics of neo-classical short-term production theory	Aspects of long-term production theory
1 In the short term, the scale of production and technology employed cannot be changed.	1 In the long term, the scale of production and technology employed can be changed.
2 In the short period, the law of diminishing marginal returns (law of variable proportions) applies.	2 In the long term, increasing returns to scale are likely to apply.
3 Factors of production (e.g. labour and capital) are imperfect substitutes for each other (i.e. the law of variable proportions) but labour-saving capital investment is possible.	3 In the long term, the law of variable proportions does not apply – no factor of production is fixed in supply.
4 Any one unit of labour (or of capital) is a perfect substitute for another unit of labour ('homogenous factors of production').	4 Units of labour and capital are imperfect substitutes for each other ('heterogeneous factors of production').
5 There are no external economies or diseconomies (i.e. the production costs and investment decisions of individual firms are mutually independent).	5 External economies and diseconomies exist (i.e. the production costs and investment decisions of individual firms are mutually interdependent).

These two tables set out elements of market and short-term production characteristics which are essential components of the full version of neo-classical international trade theory. It should be emphasised that the extent to which the neo-classical assumptions are unrealistic does not diminish their relevance to economic analysis – all theoretical structures involve a degree of abstraction or of simplification – however, it is important to bear in mind that the theoretical assumptions are not intended to be a direct representation of reality (Myint, 1965). In Chapter 8, these issues are discussed in more detail in the context of policy analysis, neo-liberalism and the 'Washington Consensus'.

the institutional structures of the global economy are required in order to achieve shared international objectives. One of the principal implications of the contrast between the approaches of the two schools of thought is that structuralist economists generally accept the place of value judgements in their analysis, usually making them explicit, while neo-liberal and neo-classical economists are considerably less inclined to accept the relevance of value judgements in 'positivist' economic analysis (and are often criticised for making their own value judgements implicit and less than transparent).

The significance of differences within a number of developing country sub-groupings and interests, and the identification of overall developing country interests is a major concern of the structuralist school of economists. The neo-liberal school of thought tends to be perceived as being more associated with the *status quo*, favouring the market position of the more powerful developed market economies and the transnational corporations which increasingly dominate the global economy. The structuralist school of thought also tends to be more sympathetic to the interests and welfare of the poorer and less privileged people in high-, middle- and low-income economies, as well as to the interests of developing countries.

While the basic neo-classical framework of trade theory is theoretically elegant, it is essentially static, and a more 'dynamic' approach incorporating increasing returns to scale, external economies, capital accumulation, technological change, institutional change, and labour force growth and enskillment can be adopted using, at its most simple level, the same basic conceptual framework as the static approach. Myint makes this point in a late 1950s article (1958), and again in a chapter produced for a World Bank collection edited by Myint in the late 1980s (1987). The significance of 'increasing returns' for economic growth in the long run has had a place in the literature for a considerable amount of time, and even Ricardo hints at this in his chapter 'On Machinery' in the third edition of his *Principles of Political Economy and Taxation* (Ricardo, 1953: Chapter XXXI). Further modification of the basic components of this 'free-trade' theory would need to allow for market imperfections (monopoly and monopsony power), product and factor differentiation, and imperfect knowledge (including information asymmetry) and the main elements of these issues have been outlined in Boxes 7.1 and 7.2.

Thirlwall sets out some robust criticisms of the static theory in his major development economics textbook making the point that 'comparative cost analysis glosses over the fact that comparative advantage may change over time' (Thirlwall, 2006: 542–4). He pursues this point further in his briefer and more specialised book on international trade and payments (Thirlwall, 2003b: 5–6). Sunanda Sen has recently extended this criticism of the 'Free Trade Paradigm' to embrace New Trade Theory, which includes more dynamic elements than those used in the basic neo-classical theory (2005: 1015–20) and which have been outlined in the long-term approach to production economics set out in Box 7.2. Sen's article is consistent with the 'North–South' approach to the explanation of uneven international (and national) development which was mentioned in this context in Chapter 2 of this book, citing articles by Krugman (1981), Ocampo (1986) and Dutt (1988), representing a criticism of the foundations of the conventional neo-classical economic theory relating to international trade.

In the real world, a variety of 'imperfections' in international markets limits the extent of factor and commodity mobility. Not the least of these is the extent of asymmetry in information and in market power which seriously impinges on the predicted outcomes from the conventional neo-classical perspective (Stiglitz, 2002: ix and 85). In recent years, there have been considerable reductions in the international barriers which inhibit the free movement of commodities and of capital through the embodiment of 'liberalisation' principles in the regulations of the GATT and the WTO (Hoekman et al., 2002: *passim*; Winham, 2005: 106–13). The significant reduction in transport costs experienced in recent decades has had the effect of reducing 'barriers to trade' (see Box 7.3) within the context of 'globalisation' which is discussed in detail in Chapter 6 of this book. Free movement of capital, of commodities and of technology is consistent with the interests of transnational corporations (TNCs), which are 'footloose' and which, inter alia, seek 'cheap labour' in international markets. The significance of heterogeneity in the labour market is clear – the interest of TNCs in ensuring the mobility of some types of labour (e.g. highly skilled technical, managerial and financial personnel, with associated guaranteeing of work permits in 'host' countries) and the immobility of other types of labour (lower-skilled workers) should be apparent. Collier outlines the factors which make many sub-Saharan African countries unattractive hosts for TNCs due to low labour productivity, partly due to characteristics of the labour force, but more significantly due to poor infrastructure and the nature of the economic and business environment in many African countries (Collier, 2000).

The question of whether there is evidence to support the proposition that international prices of factors of production have a tendency to converge in the long term is problematic. There should be no expectation that such a convergence will occur within decades rather than centuries. While there might have been some convergence in recent decades, evidence suggests that incomes have tended to polarise both between countries and within countries (Wade, 2005 and refer to the discussion of the impact of globalisation on income distribution in Chapter 6 on globalisation).

7.3 Trade liberalisation and trade strategy

The general movement away from significant and systematic trade protection, and towards trade liberalisation (i.e. the removal of barriers to trade, such as import and export taxes and import quotas) has been a feature of economic policy reform in international markets over the last three decades, and has been a major element of globalisation (refer to Chapter 6). In many respects, the process of trade liberalisation (the 'officially preferred' basis for trade policy) has – paradoxically – been associated with significant development of international regulation, nominally designed to ensure 'fair competition' and affecting both developing countries and developed industrial countries. The European Union is a prime example of a regional economic grouping which has considerably increased the freedom of movement for commodities and for factors of production within its widening boundaries. Across the globe many regional bodies have been established which aim to encourage free movement of factors and of commodities within their sub-global confines, while (in many cases) maintaining barriers to imports from outside the region (see, for example, Edwards, 1985: 224–31; Thirlwall, 2006: 523–8; Ravenhill, 2005b). These regional bodies vary in a continuum from limited trading agreements to full economic (including monetary) integration. A substantial review of the roles of trade and of trade policy in economic development is provided in a chapter by Meier (1990).

There can be little doubt about the development policy relevance of two of the main propositions associated with trade liberalisation – first, that the removal of barriers to imports which had the effect of protecting 'inefficient' domestic production of import substitutes can improve short-term consumer welfare through making cheaper imported alternatives available and, second, that the removal of trade protection from 'inefficient' domestic producers of potential exports can have the effect of improving incentives for improved efficiency and for increasing exports (Kirkpatrick, 1987: 74; Pack, 1989: 347–53; Rodrik, 1995: 2931–44; Tribe, 2005). The intention here is to focus on two specific issues – trade policy and industrialisation, and the development of exports (and of exports of manufactures in particular) from developing countries to developed industrial economies.

Neo-liberal criticism has focused on the policies (associated with a 'protectionist' approach) towards industrialisation taken by many developing countries from the 1940s to the 1970s dating from the path-breaking article by Rosenstein-Rodan (1943), which embraced the 'big-push' strategy, and extending to the approach outlined by Sutcliffe (1971: Chapter 3). A seminal contribution to the intellectual argument for the import substitution strategy for industrial development was made by Arthur Lewis in a report to the colonial government of the Gold Coast (Ghana) (Lewis, 1953). The main attack on this approach came initially in the OECD-sponsored study by Little et al. (1970) which provided damning evidence of the inefficiencies associated with government-supported development of manufacturing industry in six countries (Brazil, India, Mexico, Pakistan, the Philippines and Taiwan). This study concluded that a) trade protection was an inefficient and inconsistent means of promoting industrial development, and b) that administrative measures relating to the manufacturing sector encouraged economic rigidities which were arbitrary and costly.

Calculated effective rates of protection referred to in these studies indicated that the degree of real protection given to many industries was far higher than had ever been intended, and that this protection varied randomly between industries and countries (Little et al., 1970: 174). This OECD study can be regarded as being based on a neo-classical intellectual framework; however, many of its conclusions would be acceptable to economists within the 'structuralist' school of thought. The 'effective rate of protection' (ERP) refers to the fact that import-replacing domestic production protected by tariffs has tended to have a high requirement for imported inputs. The domestic value added (the cost of labour and of local inputs) is the contribution to national income, and it is this which is protected by the tariff. The ratio of the tariff to the domestic value added (i.e. the ERP) has often been considerably higher than the 'nominal' rate of protection provided by the tariff. For example, the arithmetic could imply that a nominal rate of tariff protection of 20 per cent would become an effective rate of protection of 100 per cent if the protected domestic economic activity had an import content which was 80 per cent of the factory-gate value of production (the ERP is explained by Thirlwall, 2006: 557–9).

The neo-liberal stance on industrial development emphasises the need to open up developing country markets to competition from imports (trade liberalisation) and to stimulate exports of manufactures from internationally competitive developing country producers to developed country markets (involving a switch from 'import substitution' to 'export promotion') (Myint, 1972: Chapter 3; Balassa, 1989). This stance is closely linked to 'getting prices right' – a major guiding principle of the neo-liberal school of thought. Import substitution, directly linked with the protectionist policies which were being phased out, was to be abandoned. In a recent book, Chang (2002) has attacked the neo-liberal approach, arguing in favour of selective trade protection of industries in developing countries. This position is controversial, since there is very wide agreement amongst economic advisers that unselective or general trade protection is an unacceptable basis for industrial development (see, for example, Wood, 2004: 934).

Import substitution has often been unfairly criticised since it is the most common form of industrial development – being 'market-led' and not necessarily requiring trade protection. Many commercially oriented industries which have been established within an import substitution context are a) fully internationally competitive, or b) not internationally competitive on the basis of production costs but transactions costs (e.g. transport costs) provide 'natural protection' of the domestic market (Tribe, 2000: 47). In addition to export promotion, other complementary industrialisation strategies include 'resource-based' and 'basic needs' approaches. These four approaches are not inconsistent with the forms of economic efficiency espoused by neo-liberal economists (refer also to Bruton, 1989).

Through focusing on the relationship between trade policy and
industrial development, economists have recently tended to lose
sight of the need for an economically coherent industrial policy for
developing countries. The 'outward-oriented' view has tended to
neglect the significance of domestic markets for manufactures, and
of exports to regional markets in neighbouring countries. The
economic literature on non-trade aspects of industrial policy, which
include alternative promotional policies (for example, the provision
of industrial estates, market studies, investment incentives and
employment incentives) is sparse: a point made by Greenaway in
his review of 'New Trade Theories' (1991).

Anne Krueger, representative of a neo-classical (or even of a neo-
liberal) view, was scathing in her criticism of the 'infant industry'
approach in her 1997 presidential address to the American Economic
Association, her line of argument closely following that of Baldwin
(1969):

> Quite aside from the unpredictability and immeasurability of the
> future time path of costs in new factories and the moral hazard
> associated with asking individual entrepreneurs to indicate how
> much protection they need, there is nothing to my knowledge in
> the literature specifying how the policy maker might instruct a
> bureaucrat to identify (much less measure) a dynamic externality if
> it were present, how an incentive-compatible mechanism might be
> devised for improving welfare, how the bureaucrat might measure
> the height of warranted protection, nor how policy makers might
> credibly commit to temporary protection.
>
> (Krueger, 1997: 12)

Krueger's emphasis on the role of the bureaucrat is perhaps misplaced,
since the process of negotiation of policy measures is shared between
bureaucrats, politicians and professional staff (including economists).
However, it is widely accepted that there are serious questions relating
to the identification of 'infants', to the *ex ante* identification of
realistic production cost trajectories over time (related to the separate
issues of capacity utilisation, learning curves and external economies
inter alia), to the specification of suitable and effective time-bound
protective (or promotional) measures (which may be trade related or
non-trade related), and the need to distinguish between protection
of products within the same trade classification (i.e. the same SITC
group).[1] Rodrik (1995: 2936) significantly makes the point that there
has been a comparative dearth of studies which test the infant industry
hypothesis directly.

Those who argue for selective and time-bound trade protection of infant industries tend to cite the fact that many of the presently developed industrial countries relied on trade protection in their early stages of manufacturing development. The suggestion is that the developed industrial countries are guilty of serious inconsistency (or hypocrisy) if having themselves attained high levels of per capita income and of development at least partially as a result of protective measures in the early stages of manufacturing development, they then deny current low-income countries the opportunity to adopt such measures (Chang, 2002). However, if Germany and the United States used trade protection to encourage the development of their manufacturing industry in the nineteenth century, it does not necessarily follow that similar policy measures are appropriate for developing countries in the late twentieth and early twenty-first centuries. Contemporary international patterns of production and trade are very different from those which existed 150 years ago, and the 'late-comers' (or 'differences in conditions') issue is very important (Myrdal, 1970: Chapter 2).

Anne Krueger's presidential address (cited above) particularly emphasised the fundamental importance of comparative advantage, and devoted considerable attention to the allocative problems created by the adoption of protective trade policies associated with the concept of 'infant industry' and the policy of import substitution from a perspective which is at the other end of the continuum to that of Chang. She accepted that the

> notion that dynamic considerations and externalities might imply that an industry, although economic, would not be established by private agents had been accepted by economists as a legitimate exception to the case for free trade since Hamilton and List . . . The premises underlying import-substitution policies were so widely accepted that developing country exceptions were even incorporated into the General Agreement on Tariffs and Trade (GATT) articles.
>
> (Krueger, 1997: 4–5)

Thus, while implicitly accepting the principle of changing comparative advantage over the long period, Krueger effectively rejects the conventional arguments for 'infant industry protection' in practice, while accepting that such protection can be justified on theoretical grounds. A similar point is made by Chang:

infant industry protection up to eight years is still allowed,
although it must be pointed out that infant industry protection was
not [italics in original] the clause invoked by countries like Korea
in protecting their infant industries under the old GATT regime.
What they usually used was the balance-of-payments clause,
which also still exists under the WTO.

(Chang, 2003b: 268–9)

There is therefore a basic dichotomy between a school of economists,
the neo-liberals, which has effectively set its face against trade
protectionism as a policy instrument designed to foster industrial
development in developing countries, and another school, the
structuralists, which has considerable reservations about the
implications of trade liberalisation for industrialisation strategies in
developing countries. For the neo-liberals, comparative advantage is
the fundamental guiding principle, but the structuralists have placed
more emphasis on the methodological limitations of static comparative
advantage theory and on the balancing of short-term and long-term
economic considerations, effectively taking the intellectual position
of 'New Trade Theory'. Another aspect of the alleged hypocrisy of
developed countries is the continued maintenance of subsidies and
other forms of trade protection, particularly for agriculture, at the
present time. The slow process of the Doha Round of trade reform –
which particularly targets developed countries – is cited as evidence
for the view that while developed countries impose trade liberalisation
on developing countries they continue to protect their own industries
(see Box 7.4). One attempt at disentangling the apparent conflict
between economists specialising in international trade and those
specialising in industrial development is represented by Tribe (2000),
with a distinction between the *protection* and *promotion* of new
industries in developing countries.

The issue of whether trade liberalisation per se contributes directly
to economic growth is controversial, and there is contradictory
analysis and evidence. A comparatively early but systematic study is
by Clarke and Kirkpatrick (1992), but a study which is widely
referred to is an analysis of long-term cross-country evidence
undertaken by Sachs and Warner (1995). This study concluded that
an 'open' trade policy positively affects economic growth. Although
the Sachs–Warner study is much quoted it has been seriously
criticised, and the particular forms of trade liberalisation which were
an integral part of the 'structural adjustment programmes' of the

1980s and 1990s cannot be uncritically regarded as leading to a beneficial effect on the economic growth of the adjusting developing countries. Rodrik and associates cast serious doubt on the conclusions of Sachs and Warner, largely due to methodological questions relating to the detailed characteristics of the database, to the operational definitions adopted, and to the analytical techniques used in the study (Rodriguez and Rodrik, 1999). Greenaway et al. (1998) record an agnostic view which echoes that of Rodriguez and Rodrik, building on an earlier review of trade theory by Greenaway and Milner (1987). This controversy is familiar in the context of discourse focused on policy-related empirical analysis and prescription. A significant part of the analytical problem is associated with the establishment of the 'counterfactual' – what would have happened in the absence of the policy change – within incremental, or 'with'/'without', analysis (see Killick et al., 1998: 19–21; Sumner and Tribe, 2008a: 146–9).

For example:

- if a policy measure has been successful in one part of the world, it does not necessarily follow that it can be transferred with equal success to another part of the world;
- complex policy adjustment takes place over a number of years with economic impacts which are only likely to take effect after a significant time-lag;
- policy reform which is not in the 'trade liberalisation' category may have economic impacts which are difficult to separate from the effects of trade liberalisation per se;
- there may be other 'economic' events which occur which affect economic performance but where these effects cannot easily be separated from the impact of trade liberalisation;
- economic data on which empirical analysis is based are uneven in quality and only become available after a significant time-lag.

The controversies surrounding access for the manufactured exports of developing countries to the markets of developed countries date back at least as far as the United Nations Development Decade of the 1960s when the United Nations Conference on Trade and Development was established.[2] Some developing country export development has involved 'final stage' assembly (for example, of electronic goods and garments) of imported materials and components, particularly associated with export processing zones with limited contributions to the domestic economy of the host country (EPZs – see, for example,

Oxfam, 2002: 181; Cypher and Dietz, 2004: 422–8). Other manufactured exports have involved further processing of locally produced raw materials, some of which were formerly exported in an unprocessed form (referred to as export-oriented 'resource-based' industrialisation). In reality, of course, it is not 'countries' which trade, but 'companies', and in this context transnational corporations (TNCs) have a particular interest in the opening up of the markets of developed countries to the exports of developing countries as TNC production and trading systems become increasingly 'globalised'. However, there has been increasing criticism of the trade policy of developed industrial economies, which restricts access for imports from developing countries while at the same time insisting that liberalisation of import restrictions by developing countries should apply to their exports (Oxfam, 2002; Binswanger and Lutz, 2003; UNDP, 2005: Chapter 4 and Box 7.4 on the Doha Round later in this chapter).

Box 7.3 gives some significant detail about the nature of 'barriers to trade' which trade liberalisation aims to reduce. It can be seen that some of these barriers are 'natural' and others are 'man-made', and it follows that the 'man-made' barriers will often be easier to reduce or to remove than 'natural' barriers. One significant barrier to trade is represented by international transport costs. Recent technological developments in sea transport, including containerisation and the increased size of ships (see Figure 7.1), have dramatically reduced transport costs thus reducing this particular 'trade barrier'.

Increasing exports from developing countries involves a degree of competition between countries within the 'developing country' grouping. The considerable expansion of manufactured exports from South-East Asian countries, and from some of the 'transition' economies which were formerly part of the Soviet bloc, implies that some of the least developed countries (particularly in sub-Saharan Africa) have a bleak prospect for the expansion of their manufactured exports to developed market economies (another form of the 'late-comers' issue). Thus, from a 'development economics' viewpoint there are two critical asymmetries – the first relating to the market power of developed industrial countries by comparison with developing countries, and the second relating to the comparative market power of various sub-groups within the developing countries category. It is the first asymmetry which is principally referred to by Stiglitz in his discussion of globalisation (2002). On the second

Box 7.3

Barriers to trade and trade liberalisation

This chapter has explained the theoretical basis behind the recent drive for trade liberalisation through trade negotiations, and through 'structural adjustment', 'economic reform' and 'trade reform'. One approach to liberalisation is a 'partial' one through the creation of regional trading agreements (RTAs). For RTAs the barriers to trade are reduced between member countries, but remain (or may be re-inforced) between member countries and the rest of the world.

'Barriers' refer to 'obstacles' to trade – and liberalisation is intended to reduce such obstacles. There are several different types of obstacle which may be the focus of liberalisation:

a) Tariffs (i.e. taxes) on imports – this increases the domestic price of imports over the landed price (c.i.f. – cost, insurance, freight). For many developing countries, the most 'efficient' way to collect tax revenue in the past was through taxes on imports – but this had the secondary effect of creating a barrier to trade and alternative forms of taxation have been widely adopted.

b) Taxes on exports – this diverts part of export revenue away from the domestic producers and towards government. Again, this was an 'efficient' way of collecting tax revenue for many developing countries in the past but it has been almost totally abandoned.

c) Quotas on imports – these exist when the physical volume of particular imported products is quantitatively limited, so that no more than that limit can imported – this restricts the supply of imports to the domestic market and would increase the domestic price of the product (depending upon market characteristics).

d) Quotas on exports – these exist when the physical volume of particular exported products is quantitatively limited so that no more than that limit can be exported – this restricts the supply of exports to the international market and would increase the world price of the product (depending upon market characteristics).

e) Transactions costs – the most obvious transactions costs are those associated with transport (including insurance etc.) so that high transport costs provide a form of 'natural protection' of domestic markets. However, transactions costs can be made artificially high through – for example – bureaucratic measures.

f) Physical requirements – many products entering into international trade are subject to technical requirements relating to health and safety legislation. It may be argued that some of the requirements imposed by developed countries on imports from developing countries are excessively severe. However, there are important phyto-sanitary conditions, for example, designed to protect consumers from potentially dangerous trade in hazardous foodstuffs (e.g. containing toxins) (refer for example to the OECD website – www.oecd.org).

7.1 Ocean-going container ship

'Panamax container ship Shenzhen Bay transiting the Panama canal. Note that some containers have been offloaded and are transported by train over the isthmus to allow acceptable draught for the ship.'

Ocean-going container ships have revolutionised international shipping, representing a significant technological change, together with the harbour installations which complement them. International shipping costs have fallen significantly over the last few decades as a result of this, reducing 'barriers to trade' in the process.

Photograph: Wikimedia – photo taken by uploader. Williamborg 04:03, 22 July 2006 (UTC) (http://commons.wikimedia.org/wiki/File:Panamax_container_ship.JPG#file)

Reproduced by permission.

asymmetry, Chhibber and Leechor (1995) have an interesting discussion about the management capacity of African countries (specifically Ghana) to respond effectively to market opportunities in a similar manner to the East Asian countries. In her detailed and important study of developing country bargaining and coalitions in the context of the GATT and of the WTO, Narlikar uses theoretical and empirical analysis to explore these asymmetries systematically (Narlikar, 2003).

7.4 Recent controversies concerning trade and development

Over the last five or six decades there have been robust exchanges within the development economics literature over international prices for the exports of developing countries. In Chapter 2, we have outlined the main features of the Prebisch–Singer thesis relating to long-term adverse trends in the terms of trade between developing countries and developed industrial countries. However, in that chapter we did not review the empirical evidence which has been presented in support of, and against, the thesis. There is a substantial amount of literature on this subject, and a good starting point is provided by two contemporary development economics textbooks. Todaro and Smith (2008: 592) show that over the period 1960 to 2005 the principal primary commodities exported by developing countries have experienced 'real' price reductions in the range of 40 to 50 per cent, although the graphs presented in their Figure 12.1 also show a considerable amount of price instability. In the seventh edition of his major textbook, Thirlwall presents evidence from the International Monetary Fund showing that between 1900 and 1990 the terms of trade for primary commodities experienced an adverse movement of around 40 to 50 per cent (2003a: 665). In the eighth edition of his textbook, Thirlwall presents a very useful summary of the debate on this issue (2006: 552–5), outlining most of the points which are included in the discussion that follows. It should be noted that the 'terms of trade' (strictly the 'net barter terms of trade') are defined as the price index for exports divided by the price index for imports. If the prices of exports are falling relative to the prices of imports, this is described as an 'adverse movement' of the terms of trade. Clearly the actual ratio of the two price indices will depend upon the statistics which are selected for the analysis – the range of commodities, the range of countries and the range of years that are included.

The details of the debate over the movement of the terms of trade between primary commodities exported by developing countries and manufactures exported by higher-income developed countries revolve around a number of issues. These include the time period selected for the analysis, differentiation between long-term trends and short-term discontinuities (e.g. the Second World War, the Korean War, and short-term international 'oil shocks'), inclusion or exclusion of specific commodity groups (particularly oil and minerals), and the precise definition of 'terms of trade' which is used (Spraos, 1980, 1985; Sapsford, 1985, 1988; Sapsford and Balasubramanyam, 1994; Sapsford and Chen, 1999). More recently it has been claimed, with robust evidence, that the Prebisch–Singer argument can be extended to developing country exports of comparatively low-technology manufactures to developed industrial countries and does not relate only to developing country exports of primary commodities (Sarkar and Singer, 1991; Lall, 2000; but see Athukorala, 1993 for a critical view). Deteriorating terms of trade for developing countries imply that part of the long-term economic growth of these countries is transferred to developed industrial countries through changes in relative prices. This is the essential point about 'the distribution of gains from trade' made in Singer's original contribution (1950). In other words, in order to achieve any given improvement in economic welfare (poverty reduction for example) developing countries have to achieve higher productivity gains with deteriorating terms of trade than without.

It is significant that a mid-1990s World Bank study of the impact of structural adjustment policies in sub-Saharan Africa calculated that for several countries deteriorating terms of trade wiped out most of the international aid transfers associated with policy reform.

> The flow of external resources has risen for four of the seven countries – Tanzania, Ghana, Kenya and Burundi, in that order – during the adjustment period. Cote d'Ivoire and Senegal have experienced declines in the ratio of net external resource transfers to GDP. Only Nigeria has had negative net flows of external resources. But even where external flows were rising, they were unable to compensate for terms of trade losses: allowing for terms of trade losses, six of the seven countries had declines in net external resource inflows, with Tanzania the lone exception.
>
> (Husain, 1994: 7)

Of course, the external aid flows which are referred to were intended to contribute to a) easing the policy reform and adjustment process, and b) the long-term development of the recipient countries, and not to compensate for terms of trade losses.

Individual commodities experience different influences on price movements in international trade, affected by a variety of factors (Newbery, 1990). For example, prices in the international coffee market are considerably influenced by the impact of frosts in Brazil (the leader in the world coffee market). International and civil warfare has been important – for example, higher prices for primary commodities during and after the Korean War in the early 1950s, and higher crude oil prices associated with crises in the Middle East in the 1970s and with the conflicts surrounding Iraq and Iran more recently.

In addition to long-term trends in the levels of international prices, the instability of prices and incomes has important implications for macroeconomic stability, for economic growth, for price incentives to commodity producers, and for incomes of domestic commodity producers in developing countries. International commodity price movements do not follow neat and symmetrical cycles (Newbold et al., 2005) and prediction of export volumes and export prices is difficult. So price stabilisation is not easy, and it is logically impossible to stabilise both prices and incomes simultaneously. Stabilisation schemes (based on buffer stocks – which are commodity specific – or buffer funds – which are not necessarily commodity specific) cannot be guaranteed to stabilise either prices or incomes. A few attempts have been made to set up international stabilisation schemes (such as the International Coffee Agreement) but conflicts of interest between individual producing countries, and between producing and consuming countries, have limited the opportunities for success in achieving stabilisation objectives (refer for more detail to Thirlwall, 2006: 559–67). Not least of the problems has been that when countries complained of 'instability' of commodity prices, they were often referring to deteriorating long-term terms of trade (Maizels, 1987, 2003).

Although a higher proportion of exports from developing countries are now manufactures, many countries are still highly dependent upon primary commodity exports (Maizels, 2003). The international financial institutions (IFIs) have led a succession of 'structural adjustment programmes' for developing countries and transition economies since the late 1970s, with advice largely being given on the assumption that each country can be treated individually and separately – an assumption which is not well founded. The phenomenon of the 'fallacy of composition' is highly relevant to the late twentieth and early twenty-first centuries in this context. The 'fallacy of composition' refers, in this context, to the fact that while individual producing countries cannot affect international prices of commodities through their production decisions, if a large enough number of individual countries take the same decision to increase production (on the assumption that the price will not be affected by their individual decisions), then the aggregate impact of increased supply will be sufficient to reduce international prices. Refer to Rodrik (1995) on developing country trade, Stiglitz (2002: 116) on the East Asian financial crisis of 1997, and Mayer (2002) on trade in labour intensive manufactures on various aspects of the 'fallacy of composition'.

It is arguable that insufficient attention has been given by the IFIs, and by the major global economic powers, to reform of the international economic structures on which all developing countries depend. Such reform, in order to make it more consistent with the priorities of the developing countries, has been given considerably less emphasis than domestic reform programmes for individual developing countries within 'structural adjustment programmes' which were given a very high priority by the IFIs.

In the 1970s, the notion of a New International Economic Order (NIEO) which would reduce the extent of perceived discrimination against developing countries in the world economy gained considerable prominence (Edwards, 1985: 231–41; Toye, 2003: 28). However, the NIEO became side-tracked following major international de-stabilisation arising from relatively high inflation, re-alignments of foreign exchange rates and changes in the role of the international financial institutions, so that the concerns of the major economic powers became focused on the problems of the developed industrial countries rather than on those of the developing countries. It was at around this time that the General Agreement on Tariffs and Trade (GATT) was re-invented as the World Trade Organization (WTO) (Cernat et al., 2002; Group of 77 and China, 2001; Hoekman et al., 2002: passim; Winham, 2005). By the early 1980s, the NIEO had been side-lined (Fortin, 1988: 71; Evans, 1989: 1290–2; Stiglitz, 2002: 13 and 65). It was followed by the Washington Consensus (Williamson, 1994) representing resurgence of neo-liberalism over structuralism as the guiding principle for economic policy (Toye, 2003: 31–2) and this will be discussed in more detail in Chapter 8.

In the last two decades the international economic agenda has changed considerably due to a number of significant structural changes. One of these changes has been the accession of the Peoples' Republic of China into the WTO in 2001 (see, for example, Shafaeddin, 2002a and 2002b) and its increasingly influential role in the international economy.

Other changes have been associated with the continued unravelling of the colonial legacy. Over a long period, there were significant efforts to add a 'developing country' dimension to the structure of the international institutions which had an 'over-seeing' role within international trading structures. Thus, the original purpose of the General Agreement on Tariffs and Trade (GATT) was to provide a framework within which the major trading nations could operate. During the colonial period the interests of the developing countries tended to be regarded as peripheral and were represented by the colonial powers within international institutions. After the gaining of independence by the former colonies during the latter part of the 1950s and into the 1960s, the newly independent nation states clearly had a range of issues which needed to be addressed directly rather than through the 'good offices' of the former colonial powers. Thus, ensuring better access for developing country exports to developed industrial country markets, and ensuring a 'fairer' overall international market position for developing countries, became a more substantial part of the international 'agenda'. Initially, this occurred through what was referred to as the 'Uruguay Round'. In more recent years, following the replacement of the GATT by the World Trade Organisation (WTO), this role has been replaced by the 'Doha Round'. Box 7.4 outlines the main issues associated with the Doha Round.

Another aspect of this unravelling of the post-colonial institutional matrix concerns the association between the EU and the African, Caribbean and Pacific group of countries (ACP). Most of the ACP countries were formerly British and French colonies, and many of the special trading relationships which had existed during the colonial period were carried over into the post-colonial period and were further developed – for example, through the ACP arrangements. Box 7.5 outlines the latest developments in these relationships, following the WTO's rejection of many of the provisions of the ACP agreements. These changes are clearly of the utmost significance for many developing countries.

Box 7.4

The Doha Round of trade negotiations

An attempt was made to address the international trade concerns of developing countries with the launch of the Doha Development Agenda (DDA) in 2001, although as of early 2009 these negotiations have not been completed successfully. The Doha Declaration (see www.wto.org) provided a mandate for negotiations on a range of issues including:

- agriculture: the establishment of a fair and market-oriented trading system, including better market access, the phasing out of export subsidies and substantial reductions in domestic support that distorts trade;
- services: the General Agreement on Trade in Services (GATS) commits member governments to enter into negotiations to liberalise progressively trade in services;
- market access for non-agricultural goods; although average tariffs are at low levels, certain tariffs continue to restrict trade, especially exports from developing countries; 'tariff peaks' (relatively high tariffs on 'sensitive' products) and 'tariff escalation' (higher import duties are charged on semi-finished products than on raw materials and are higher still on finished products) remain issues that need to be resolved;
- trade-related aspects of intellectual property rights (TRIPS); these relate in part to issues of public health and access to medicines, geographical indications (place names used to indicate products with particular characteristics – Scotch whisky and Parma ham, for example) and biodiversity;
- other issues include ensuring that regional trade agreements remain compatible with WTO rules.

Negotiations take place in the Trade Negotiations Committee and its subsidiaries, and the original Doha Declaration has been refined at a number of meetings including Cancun, Mexico (2003), Geneva (2004, 2006, 2008), Hong Kong (2005), Paris (2005) and Potsdam, 2007. The negotiations collapsed in July 2008 in Geneva. The basic conflict is between the EU, the USA and Japan on the one side and Brazil, Russia, India, China and South Africa (BRICS) on the other, failing to reach agreement on agricultural subsidies, industrial tariffs, non-tariff barriers and trade in services. In Geneva, it was the USA versus India and China arguing over Special Safeguard Measures (in effect, the latter two economies refusing to liberalise their domestic markets for subsidised imports from the USA) that led to the breakdown in negotiations, although the EU and the USA also disagree on the issue of agricultural subsidies. Cotton subsidies by the USA have been a particular bone of contention to West and Central African cotton producers. Colman (2007: 98–100) looks in detail at the conflicts between the EU Common Agricultural Policy (CAP) and the Doha Round negotiations.

continued

As of early 2009, the WTO (and some developing economies, for example, Brazil) have indicated their wish for agreement to be reached on the DDA. The WTO in particular sees the successful conclusion of the DDA as an important input into the solution of the 2009 global economic crisis. Not everyone shares this view, however.

Box 7.5

Economic Partnership Agreements (EPAs)

The European Union (EU) has had a variety of arrangements with the African, Caribbean and Pacific (ACP) countries since 1963, culminating in the Cotonou Agreement of 2000 (Nixson, 2007: 340). With respect to trade, the objectives of the Agreement were to accelerate the smooth and gradual integration of the ACP states into the global economy, to help them rise to the challenges of globalisation and adopt the new conditions of international trade, as being defined by the World Trade Organisation (WTO), and to enhance their production, supply, and trading capacities and improve their competitiveness. But the trade regime enshrined in the various conventions was not compatible with the WTO's rules and the preferential treatment extended on a non-reciprocal basis to ACP countries was to be phased out over a ten-year transitional period (to end 2007).

The EU's preferred solution to this problem was the creation of Economic Partnership Agreements (EPAs) between the EU and ACP countries that would require the latter, among other things, to liberalise their imports from the EU in order to comply with WTO Article 24. If an agreement on the EPA was not reached, the ACP states risked losing their privileged status in the markets of EU member states.

The negotiation of EPAs has proved to be controversial and complex. Some progress has been made, however, and Interim EPAs (IEPAs) have been negotiated with ACP states in western, central, eastern and southern Africa, the Caribbean and the Pacific. But deadlines have been missed. For example, Papua New Guinea and Fiji in the Pacific have initialled the Interim EPA to protect market access for exports of tuna and sugar to Europe. But both countries are unhappy with the terms of the Agreement and take the view that rules on export taxes, protection of infant industry and most favoured nation (MFN) treatment should be revised.

Critics of EPAs argue that further trade liberalisation, even if it benefits ACP countries in the long run, will impose short-run adjustment costs related to implementation and restructuring, will lead to the loss of fiscal revenue (arising from the reduction or abolition of import duties) and the loss of autonomy in designing and implementing national development policies. It appears that 2009 will be a crucial year for the conclusion of the negotiations establishing EPAs.

Significantly, negotiations relating to the EPAs are the responsibility of the EU's Directorate General for Trade while the Cotonou Agreement (relating to the ACP states) was the responsibility of the Directorate General for Development. This change subtly (or perhaps unsubtly) alters the balance of priorities and concerns within the negotiations.

Some sources for further information about the EPAs have been given at the end of this chapter.

7.5 Summary

- The conventional economic theory relating to international trade is based on gains from trade arising from the 'law of comparative advantage', associated with differences in factor endowment and with converging relative factor prices.
- The adoption of comparative advantage as a basis for economic policy in developing countries is highly controversial because the law of comparative advantage relates to the short run, while in the longer run comparative advantage changes due to economies of scale, technological change and external economies.
- The protagonists within the economics profession differ over the extent to which comparative advantage should be the main basis of policy and can be characterised as neo-classical (conventional – in favour) and structuralist (heterodox – against) economists. There has been a tendency for each side to question the 'economic literacy' of the other side.
- The neo-classical economic approach is often linked directly with the 'neo-liberal' ideological stance of conservative politicians. However, it is necessary to distinguish between theoretical and systematic positions within the economics profession and ideological positions within the political world.
- Trade liberalisation (the reduction of barriers to trade) has been adopted as a 'default' policy approach by most international institutions in recent decades. The advantages of this approach for developing countries is also the subject of dispute within the economics profession depending on both the theoretical position taken and the interpretation of empirical analysis.

● The question of whether the terms of trade between primary commodities and low technology manufactures exported from developing countries relative to imports of manufactures from developed industrial countries is another area of economic dispute. This dispute revolves around the interpretation of empirical data.

● There have been significant structural changes within the international economy over the last few decades, and these changes continue with the Doha Round of trade negotiations and with the EU's espousal of Economic Partnership Agreements for example.

Questions for discussion

1 Why is there a dispute over the role of the law of comparative advantage as the basis for economic policy relating to international trade for developing countries?

2 What are the main differences in perspective between the conventional theory of international trade and 'new trade theory'?

3 Why has there been significant controversy over the role of liberalisation as the guideline for the trade policy reform of developing countries?

4 Why is there no widely agreed interpretation of empirical data for the last few decades on the terms of trade between developing countries and developed industrial countries?

5 International reform of trade policy through the Doha Round is nominally intended to improve the position of developing countries in international trade. Why have these negotiations been so protracted and difficult?

Suggested further reading

Chang, H.-J. (ed.) 2003a. *Rethinking Development Economics*. London: Anthem Press, Chapter 12.

Cypher, J. and Dietz, J. 2004. *The Process of Economic Development* (2nd edn). London: Routledge, Chapter 9.

Thirlwall, A.P. 2006. *Growth and Development with Special Reference to Developing Countries* (8th edn). Houndmills, Basingstoke: Palgrave Macmillan, Chapter 16.

Todaro, M. and Smith, S. 2008. *Economic Development* (10th edn). London: Pearson Addison Wesley, Chapter 12.

Useful websites

The Overseas Development Institute and the Institute of Development Studies have a number of very helpful background documents relating to the Doha Round and to the EPAs which have been outlined in Boxes 7.4 and 7.5 (www.odi.org.uk and www.ids.ac.uk). The European Union has a newsletter relating to trade and development which can be found at www.acp-eu-trade.org/newsletter.

On the official aspects of international trade regulation and negotiations the World Trade Organization, the United Nations Conference on Trade and Development, and the Development Assistance Committee of the Organisation for Economic Co-operation and Development are essential sources (www.wto.org; www.unctad.org and www.oecd.org/dac).

Economic concepts used in this chapter

Comparative advantage/specialisation/gains from trade/free trade
Factor endowment/factor price equalisation
International exchange rates/terms of trade/'real' (constant) prices
Economic integration/monetary integration
Short-term comparative statics/diminishing marginal returns/law of variable proportions/production possibility curve/efficiency frontier
Rate of exchange/relative costs/prices/trade-off/opportunity cost
Investment/economies of scale (increasing returns to scale)/technology change/external economies and diseconomies/capital intensive/labour intensive
Nature of production/perfect competition/atomistic competition
Market structures/perfect markets/asymmetric markets/imperfect markets/oligopoly/monopoly/oligopsony/market failure/market power/fallacy of composition
Neo-classical assumptions/marginal analysis/neo-liberalism
Structuralist/structuralism/heterodox economics
Value judgements/positivist economics/theory of second best
Footloose/labour market heterogeneity (segmented labour markets)
Trade liberalisation/barriers to trade/'fair' competition/effective rate of protection/nominal rate of protection/domestic value added
Import taxes (tariffs) and export taxes/export and import quotas (quantitative controls)/transactions costs/natural protection

- Economic policy reform/IFIs (international financial institutions)/NIEO (New International Economic Order)/Washington Consensus
- Import substitution/export oriented (promotion)/effective rate of protection/infant industries/final stage assembly
- GATT (General Agreement on Tariffs and Trade)/WTO (World Trade Organization)/Uruguay Round/Doha Round/MFN (most favoured nation)/SITC (Standard International Trade Classification)
- Counterfactual (incremental – 'with/without' – analysis)

Notes

1 SITC refers to the Standard International Trade Classification which categorises commodities systematically.

2 The UNCTAD was established in 1964 in order to promote 'the development-friendly integration of developing countries into the world economy' (UNCTAD, 2009a).

8 Economics and development policy

8.1 Introduction

There has been considerable concern within development studies over the last few decades with the policy 'conditionality' associated with the aid programmes and policy advice of the World Bank, the International Monetary Fund (IMF) and other members of the international aid community. Much of this concern has related to the extent to which the policy formulation process was being taken out of the hands of developing country governments and other indigenous institutions and the questions of 'country ownership', 'donor driven' aid programmes, 'leverage' and associated concepts of neo-colonialism, have been to the fore (Hayter, 1971; Killick et al., 1998). This could be regarded as a 'political economy' dimension of development economics and of development studies. The OECD has a page on its website which defines its 'institutional' interpretation of 'political economy' (OECD, 2009), but this definition is significantly different to the concept of the 'political economy of capitalism' embraced by writers such as Fine (Fine, 2002, 2006b). Van Waeyenberge (2006) has an extensive discussion of the policy formulation process associated with structural adjustment in the context of concepts of 'political economy'.

Together the World Bank and the IMF are essentially the 'international financial institutions' (IFIs). Broadly speaking, at the individual country and international levels, the IMF takes responsibility for overseeing macroeconomic management issues, including the balance of payments (and the foreign exchange rate), aggregate demand (levels of consumption, savings, investment, public expenditure, etc.) and monetary policy (rates of interest, credit creation and supervision of the financial sector). The World Bank takes responsibility for overseeing more microeconomic issues including public policy and sectoral issues. In practice, there is more overlap than this division of responsibilities might suggest.

In the context of macroeconomic management, the policy 'conditionality' developed by the IFIs, being central to their 'Structural Adjustment Programmes', has insisted on the exercise of constraint over the aggregate level of public expenditure and has emphasised the need to reduce public expenditure as a proportion of national income. In the context of microeconomic management, the policy conditionality has insisted on market liberalisation, including the reduction of market controls, the removal of subsidies and the privatisation of many public bodies (including agricultural marketing and economic infrastructure such as public utilities). The microeconomic approach to development policy also concerns 'efficiency' issues, which amounts to getting the most out of the available resources and to achieving policy objectives at least cost. Particularly valuable reviews of structural adjustment experience are by Taylor (1988), Mosley et al. (1995), Mohan et al. (2000) and Vreeland (2003, 2007).

One of the main objectives of this chapter is to distinguish between those elements of 'conditionality' which might be regarded as essential parts of prudent economic governance and the achievement of consensual development objectives, and others which are less essential and might truly be regarded as being based on ideological preconceptions which it is inappropriate for external bodies to impose on independent nation states.

Following this brief introduction, the second and third sections of the chapter will focus on macroeconomic and microeconomic issues respectively in the management of developing country economies and the achievement of development policy objectives. Although there is no completely clear distinction between those parts of economics which relate to macroeconomics and to microeconomics, this division of the discussion is convenient. The fourth section outlines a critical approach to neo-liberalism and the Washington Consensus, which has been so influential in the last few decades. The fifth section discusses a number of key issues in the management of international aid, and the sixth section consists of a summary of the main points in the chapter.

Many of the issues raised in this chapter are also relevant to the 'transition' economies of eastern Europe, central Asia, and South-East Asia which have been concerned with a switch from dependence upon economic management based on 'controls' to one dependent upon 'markets' (World Bank, 1996). This switch is directly related to the issue of market liberalisation. The content and approach of the chapter is intended to address policy economics in the context of international development and poverty reduction in particular.

8.2 Macroeconomic policy and the management of developing economies

Significant improvements in the performance of public services such as health, education and water/sanitation – which have a clear link to poverty reduction strategies – require considerable investment, and international aid has responded strongly to this logic. International aid flows have to a large extent been 'earmarked' for these public services, reflecting policy priorities of both 'donors' and 'recipients' (i.e. 'development partners' in contemporary terminology). However, there is an important 'accelerator' effect, so that increased public investment expenditure financed through international aid programmes requires matching increases in domestically financed recurrent expenditure in circumstances where the governments of many developing countries experience major constraints due to difficulty in generating additional tax revenue. Government tax revenue can, of course, be supplemented through increased domestic borrowing and the running of a government deficit. Domestic borrowing by government may occur through diverting funds from the private sector (either households or enterprises) into the public sector, or through credit creation (borrowing from the banking system). If funds are diverted, this may involve switching from investment expenditure into recurrent expenditure, perhaps reducing savings and investment in the economy as a whole. Some international aid institutions are now prepared to commit funds not only to finance investment and 'technical assistance' (as has been traditional) but also to support recurrent expenditure by developing country governments (such as for teachers' salaries).

For most of the period since 1945, international aid has tended to support only investment and technical assistance – for example, funding the building of schools and the provision of expatriate specialist skills. Aid has tended to support specific development projects, and sometimes broader programmes, which have been selected, appraised (and sometimes designed) by the international aid institutions. More recently, aid institutions have loosened the aid management system so that for some countries funding can be provided either as sector, or as general budget, support with the recipient government being responsible for the overall management of expenditure at a project or programme level (IDD, 2006).

Expansion of public expenditure based on budget deficits, credit creation or external inflows (international aid) can lead to internal macroeconomic imbalance (excess demand) and to inflation. It is therefore necessary to balance increases in government expenditure which aim to achieve desirable improvements in public service delivery against macroeconomic prudence. It should be noted that these domestic financial factors are only part of the explanation for the occurrence of inflation, and that there are other more structural factors which need to be accounted for, including international 'contagion'. International 'contagion' refers to the ease of transfer of, particularly, inflation between countries. The analogy with contagious diseases is intentional. In addition, there have been fears that significant inflows of foreign exchange through international aid can be associated with 'Dutch Disease' symptoms leading to appreciation of the foreign exchange rate, reduced competitiveness of exports and increased incentives to import (see Box 8.1).

Another concern of many economists over increases in public expenditure relates to 'crowding out'. This refers to the fact that higher levels of public expenditure funded through increased government borrowing can reduce the extent to which the private sector can borrow from the financial system in order to fund directly productive activities, many of which would be directly linked to economic growth. Domestic credit to the private sector may be short term in nature (to fund private companies' working capital), medium term (to fund investment which has a fairly rapid 'pay back' period), or long term (to fund mortgages relating to house purchase, or

Box 8.1

The impact of the 'Dutch Disease' on the domestic economy

The concept of 'Dutch Disease' refers to the economic impact of significant inflows of foreign exchange and is explained in Cypher and Dietz's book (2004: 319–21). The origins of the term date back to the impact of the large export revenues earned by the Dutch economy through the sale of hydrocarbon products on the international market in the early 1960s (Gylfason, 2008). While higher levels of foreign exchange inflows are usually to be welcomed, sometimes they represent a mixed blessing. Where foreign exchange rates are largely determined by the supply and demand of foreign currencies (i.e. a 'floating' exchange rate), an 'excess' supply of foreign exchange (say euros) represents increased demand for the domestic currency (say Nigerian naira – as occurred through the huge increase in Nigerian oil exports some years ago). The increased demand for naira increases the international price – so that, as a hypothetical example, the exchange rate rises from €1.00 equals ₦10 to €1.20 equals ₦10. This is referred to as an appreciation of the value of the naira. If international euro prices for imports and domestic naira prices for exports remain the same, then the appreciation of the naira's international value would have the effect of decreasing the naira value of Nigerian imports from euro-zone countries (stimulating the demand for imports), and increasing the euro value of Nigerian exports (choking off demand for exports).

Another way of looking at this is that, after the appreciation of the naira, for any given euro value of Nigerian cocoa exports domestic suppliers would receive fewer naira – making cocoa production for export less attractive. Import substituting domestic manufacturing industries would benefit from the reduced price of imported capital equipment and other essential inputs, but the greater the proportion of Nigerian value added (i.e. through the use of domestic inputs and labour) the less competitive these industries would then be relative to competing imports. At the same time, Nigerian importers could maintain their naira prices to customers at the same level as before the currency appreciation and reap higher profits through increased imports from the, now cheaper, euro-zone sources of commodities.

The 'Dutch Disease' would therefore be regarded as having a 'perverse' and undesirable effect on economic incentives through the price system. In this Nigerian case, a wide range of producers would find that their market position was being challenged – through the 'double-whammy' of reduced demand for exports, and increased competition from imports.

continued

'Dutch Disease' is relevant in other cases which are of interest to development studies specialists. First, the effects on the foreign exchange rate of significant increases in inflows of foreign exchange through international aid is likely to have a depressing effect on incentives for export production and for import substituting domestic production. Second, one of the effects of the Asian Financial Crisis of 1997 was through the impact of large financial inflows on the exchange rates of the countries concerned (refer to the discussion of the 1997 Financial Crisis in Chapter 6) (Stiglitz, 2002: Chapter 4).

infrastructure investment). If the government increases its borrowing from the financial system, this potentially reduces the amount of borrowing which can be undertaken by the private sector – which 'crowds out' private investment.

'Crowding out' through government 'capture' of financial resources which might otherwise have been used to fund private sector investment has been a major concern of development economists specialising in macroeconomic management since the concept was raised by McKinnon (in the context of 'financial repression') in the early 1970s (McKinnon, 1973: Chapter 7; Dornbusch et al. 2004: 279–84). A distinction needs to be made between government borrowing which competes for existing loanable funds available from the financial system, and that which draws on 'credit creation' through a 'loosening' of credit restraints and which amounts to the 'printing of money'. It is the first of these which can be linked to the issue of 'crowding out' – public sector expenditure replacing private expenditure, while the second is linked more directly to concerns over 'excess demand' and the danger of higher rates of price inflation. Other economists (for example, Taylor, 1988: 36 and 152) have emphasised 'crowding in', the positive impact of, for example, government investment on infrastructure or of other government expenditures which stimulate private investment and private sector economic activity.

The link between higher levels of public investment and increased demand for recurrent expenditure has led to the abandonment of Public Investment Programmes (focusing only on investment expenditure – sometimes referred to as being 'below the line'), and the adoption of Medium Term Expenditure Frameworks (MTEFs) which explicitly encompass forward planning of both investment and recurrent expenditure. The separation of planning for public investment from planning for recurrent expenditure had encouraged a lack of coordination between the two. The crucial difference between 'investment' expenditure and 'recurrent' expenditure is that the former is a 'one-off' payment which increases the size of the capital stock (such as the building of a school), while the latter is a continuing commitment (such as teachers' salaries). While investment expenditure can be adjusted up or down relatively easily, recurrent expenditure tends to be very inflexible in the short term. Increasing the number of school places through an investment programme involves a continuing (into perpetuity) commitment to pay for teacher salaries.

The MTEF approach offers a hope for a more coherent approach to the management of public expenditure as a whole (World Bank, 1998a; Klugman, 2002: Chapter 6). The World Bank has a macroeconomic projection model known as the 'Revised Minimum Standard Model' (RMSM – World Bank, 1998b) which has been developed over the years and which is widely used as a basis for projecting major macroeconomic variables as a basis – inter alia – for estimating future tax revenues.

8.3 Microeconomics and development policy

The role of microeconomics in providing a theoretical basis for a systematic approach to development policy raises a key issue of contention between neo-classical and structuralist/heterodox economists. Associated with this is the important distinction between the principles of neo-classical economic theory and the tenets of neo-liberalism. Essentially, the neo-classical approach relates to a specific system of economic theory involving explicit assumptions (which may or may not be realistic) and using marginal analysis. Neo-liberals are those who propose an approach to policy design and implementation based on the principles of neo-classical economics with the key condition that the neo-classical assumptions hold in the real world, that is, assuming that 'markets work'. This can be described as the 'free market' approach. The concept of the 'free market', as espoused by the neo-liberals, has often been confused with the conditions (or assumptions) of the 'perfect market', which is a critical element of neo-classical economic theory. In the real world, 'perfect markets' do not exist (refer to Box 7.2 in Chapter 7 for a comparison between the assumptions on which perfect markets are based and the conditions which apply in the real world). The neo-liberals have been criticised on the grounds that they are either naïve (unlikely) or undertake deliberate misrepresentation (more likely) in this respect. The neo-classical economists do not, of course, claim that their theories describe the real world, but they do claim that their theories are relevant to an understanding of how the economy works (Myint, 1965; Tribe, 2006).

One of the issues associated with 'policy reform' and 'structural adjustment' has been the adoption of 'full cost recovery' principles for the pricing of, and charging for, public services. 'Full cost recovery' means that the institutions responsible for the provision of public services aim to generate revenues sufficient to cover all costs, including capital costs. This has often involved significant increases in the prices of public utilities (quite a few of which have been privatised) such as water supply and sanitation (particularly in the 'modern sector'), telecommunications and electricity. It has also involved difficult questions around charging for medical services and for education (the latter of which has a high recurrent expenditure component due to the significance of teacher salaries). Although the cost recovery approach (in many cases linked with privatisation) has led to a notable improvement in the finances of public utilities, it has also often had a significant impact on equity, making essential services less accessible to lower-income groups because of the charges which have to be paid in order to secure access, and this impact has perhaps been more serious for rural than for urban areas.

Associated with this is the role of subsidies in development policy, leading to the exploration of a number of microeconomic issues in a development context. Subsidies can take a number of different forms including:

- the *ex post* covering of deficits generated by public bodies (one of the justifications for privatisation);
- reduction of market prices of essential commodities and services for low-income groups;
- reduction of the prices of essential production inputs and stimulation of production (e.g. agricultural production);
- achievement of environmental objectives.

The economic arguments for subsidies (and for some forms of taxation) are also associated with 'market failure' which will be explained a little later in this chapter.

Historically, many public utilities in developing countries were run at a deficit in the public sector, with a need for 'subventions' from government finances at year-end to cover deficits. These 'subventions' tended to be unplanned, meaning that the percentage reduction in the market price was not known in advance (which would have been the case for a planned and targeted subsidy) and that their impact on government expenditure could not be known in advance and so tended to be 'random'. The reasons for the deficits included a lack of financial discipline, often linked to the perceived political difficulty of increasing prices for public services such as water supply, electricity supply and telephone calls. The adoption of full-cost recovery and privatisation has reduced the impact of these subsidies on government finances, at the expense of higher consumer charges. The arguments in favour of privatisation have included the view that only through the imposition of financial discipline represented by private sector operation is the stimulus for both short-term and long-term efficiency sufficiently strong to secure financial sustainability. It can also be argued that privatisation with international involvement has been a means of securing stronger management capacity and technology transfer as a basis for the improvement of services and of financial performance. Readers wishing to explore arguments and experiences relating to privatisation in developing countries more fully are referred to Cook and Kirkpatrick (2003) and to chapters in Parker and Saal (2005).

Subsidies which have welfare or equity objectives have often been suggested as part of a poverty reduction strategy related to essential services and commodities such as housing and foodstuffs. However, it cannot be assumed that subsidies are an effective policy instrument for poverty reduction, or that policy instruments which are effective in developed industrial countries will have similar impacts in developing countries. For both housing and foodstuffs (which tend to account for a high proportion of household expenditure in urban areas of developing countries), subsidies can often only be adopted where products or services are provided directly by the public sector. Since both housing and foodstuffs are typically provided from within the private sector – and this has been increasingly so since the adoption of structural adjustment/policy reform in the last two or three decades

– the effectiveness of subsidies in achieving poverty reduction objectives is likely to be limited. The impact of subsidies in achieving poverty reduction objectives through the provision of housing and foodstuffs would be uneven if they reach only a small proportion of the poor. Experience also suggests that subsidies intended to benefit the poor are often 'captured', significantly reducing the policy impact. Such 'capture' may be through direct channelling of benefits to middle- and higher-income groups which are not intended beneficiaries, or may involve the 'selling on' of goods and services at higher open market prices so that lower-income groups are able to 'capture' the 'capital gain' rather than the direct benefit which was intended (i.e. they receive money rather than the goods and services which were intended).

Subsidies can also be related to the encouragement of production (e.g. through reducing the cost of fertilisers and other farm inputs – including perhaps irrigation water – to producers). In some cases, this type of subsidy has achieved conspicuous success but, in other cases, diversion problems have arisen as with housing and foodstuffs subsidies. Agricultural 'production' subsidies have sometimes been linked to the production of specific products which may, for example, be targeted at export markets for the earning of scarce foreign exchange. Farmers can easily re-direct subsidised inputs towards the production of foodstuffs (which they may regard as being of a higher priority) or other products, or they may be sold on for other uses with the subsidy being converted into cash. Again, there can be a contrast between the intended and actual outcomes from subsidy programmes. However, Korea represents an example of how production-related subsidies directed at manufacturing sector, and export, development can successfully achieve policy objectives if carefully targeted in a 'time-bound' manner (Amsden, 1992). Amsden emphasises the importance of reciprocity, meaning that the recipient of subsidies has obligations or responsibilities to the government which if not met would lead to the withdrawal of the subsidy and the application of penalties (if, for example, export targets were not met). Box 8.2 aims to set out a number of the main economic considerations relevant to the analysis of the role of subsidies in achieving policy objectives.

There are two other issues which make subsidies comparatively unattractive policy instruments for the achievement of development policy objectives. The first is that many developing countries have

Box 8.2

The role of subsidies in principle

This box aims to explain some of the fundamental economic principles involved with subsidies. The intention of subsidies is to provide goods and services to a market at a price which is lower than cost. This stimulates demand for the particular good or service concerned in a way which is illustrated in the following diagram.

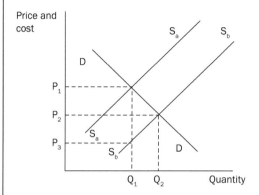

The supply curve S_aS_a shows the quantity of the good or service supplied (horizontal axis) at a market price (vertical axis). It is usually the case that the 'supply price' is directly related to the unit cost of production. The demand curve DD shows the quantity of the good or service demanded (horizontal axis) at a market price (vertical axis). P_1Q_1 shows the market clearing price/quantity relationship for these market conditions.

The supply curve S_bS_b shows the quantity supplied to the market after the application of a subsidy (which is indicated by difference between P_1 and P_3). The subsidy does not affect the position or slope of the demand curve, and the post-subsidy market clearing price/quantity relationship is shown by P_2Q_2. The effect of the subsidy is to increase the quantity demanded from Q_1 to Q_2 and to reduce the market price from P_1 to P_2. The effect of the subsidy is to reduce the market price by less than the level of the subsidy. The outcome from this hypothetical example depends upon the interaction between the slope (demand and supply elasticities) and position (level) of the respective demand and supply curves. An extension of this explanation can deal with the outcome from taxes such as those intended to discourage demand/ consumption of a good or service, perhaps for social or environmental reasons. The tax is a subsidy in reverse (or a subsidy is a tax in reverse).

continued

> Note that the administration of subsidies and taxes involves costs for both governments and private sector/civil society institutions which are usually referred to as 'transactions costs'. Fiscal neutrality (i.e. subsidies and taxes which have no effect on the aggregate levels of government revenue and expenditure) needs to take account of such transactions costs. In national income accounting terms, subsidies are defined as 'transfer payments' – meaning that, like taxes and social security benefits, there are no goods or services provided in return for the payment.

experienced considerable difficulty in raising sufficient tax revenue for the funding of the wide range of activities for which they are responsible. The commitment of a significant proportion of government revenue to subsidy programmes from which the outcomes and impacts are uncertain (or perverse) can be regarded as dysfunctional. There are alternative policy instruments which can be used for the achievement of intended outcomes and impacts. Second, ensuring that subsidy programmes have a reasonable chance of achieving at least part of their objectives can involve significant transactions costs, but even then the outcomes and impacts may not be assured. The World Bank's Poverty Reduction Strategy Papers (PRSP) Sourcebook (Klugman, 2002) discusses the role and impact of subsidies in the context of several dimensions of policy formulation – for example, in Chapters 2 (Inequality and Social Welfare), 6 (Public Spending) and 11 (Environment). This discussion of subsidies has attempted to avoid ideological issues – a difficult task. However, it is necessary to be aware that neo-liberals (as adherents of 'free-markets') tend to take the view that subsidies are an inappropriate modification of market characteristics.

Discussion about subsidies as an instrument for the achievement of policy objectives within 'markets' is often focused on the means of delivering outcomes and impacts. Subsidies are one of a range of policy instruments which might be used in order to achieve specific objectives. There are two important lessons from this approach. The first is that subsidies can be used to 'correct' for 'market failure', and the second is that in the analysis of policy measures (including the use of subsidies), the 'incremental approach' should be used. The 'incremental approach' involves a comparison between the outcome 'with the policy' and the outcome 'without the policy' – comparison of 'with' and 'without' – and the 'without' case is known as the 'counterfactual'. The 'incremental' approach is described in Sumner and Tribe (2008a: Chapter 6), Killick et al. (1998: Chapter 2) and Vreeland (2003: Chapter 5; 2007: Chapter 4).

The issue of 'market failure' has had much prominence in the development economics literature, and was a major element in the justification for economic development planning in the 1960s and 1970s and for government intervention in general (Thirlwall, 2006: Chapter 9; Meier and Rauch, 2005: *passim*). Most development economists take the view that while 'market failure' is widespread in all countries, in developing countries it is a more common feature than it is in developed market economies. Perhaps the clearest brief explanation of the nature of market failure has been provided by Meier in the sixth edition of his major text, *Leading Issues in Economic Development*. For Meier, market failure occurs when:

1) The market does not function properly – the case of market imperfections,
2) The market result is incorrect – the case of externalities,
3) No market exists for the relevant activity – the case of public goods,
4) The market yields undesirable results in terms of objectives other than resource allocation.

(Meier, 1995: 540)

Market imperfections exist when the conditions (or assumptions) of the theoretical 'perfect market' do not apply. The most important of these conditions are:

i) large numbers of buyers and sellers so that no single buyer or seller has control over a significant part of the market;
ii) freedom of buyers and sellers to enter and exit the market;
iii) perfect market information for both buyers and sellers;
iv) perfect foresight of future events relevant to the market.

The assumptions associated with economists' definitions of both 'perfect markets' and 'perfect competition' have been set out in Box 7.2 in Chapter 7.

One graphic, but probably apocryphal, example of 'externalities' is where the long-standing owner of derelict land in a city erects a large sign expressing thanks to neighbours for all of their actions over the years which have resulted in a substantial increase in the market value of the land without any contribution having been provided by the land owner. Externalities (also referred to as external economies and diseconomies) exist when any institution (which might be a firm, a household or an individual) within a market imposes costs on (or creates benefits for) another institution but where the 'originating' institution does not bear the cost (or receive the benefit) of their actions through the market.

Public goods are those which are freely accessed without payment (i.e. shared goods or services – a clear example is common land, where a major issue is that of who takes responsibility for organising and paying for care and maintenance, and another example is the atmosphere). A common misinterpretation of the term 'public goods' is that they are goods and services provided by the public sector. As a technical economic term, 'public goods' have the very precise meaning which has been given in the text.

Box 8.3 reproduces a definition of market failure taken from *The Economist* website. This not only illustrates the quality of the online glossary of economic terms available from that source, but also gives a guide to the widespread concern with the interpretation of this important economic concept.

Joseph Stiglitz (a former chief economist at the World Bank) was awarded the Nobel Prize in Economics for his joint work on the issue of asymmetric information in markets, and he has also written on the policy implications of market failure. He has taken the view that rather than being the exception, market failure is the rule. This means that market failure is all around us, and that governments will have difficulty addressing all instances of market failure. The implication is then that government interventions should be limited to a) where the market failure is 'large', b) where policy objectives are likely to be achieved successfully, and c) the outcome of the intervention should enhance 'welfare' (based on Stiglitz, 1989: 38–9). Many economists – particularly those of a neo-liberal predisposition – are very suspicious of the concept of market failure as a justification for government policy intervention, emphasising instead the role of governments in creating market failures through misguided and/or ineffectual interventions – the case of 'government failure'.

There are various types of government intervention which can 'improve' outcomes as compared by those provided by the market in the absence of the intervention:

Box 8.3

The Economist *magazine on market failure*

'When a market left to itself does not allocate resources efficiently. Interventionist politicians usually allege *market failure* to justify their interventions. Economists have identified four main sorts or causes of *market failure*.

- The abuse of *market power*, which can occur whenever a single buyer or seller can exert significant influence over *prices* or *output* (see *monopoly* and *monopsony*).
- *Externalities* – when the market does not take into account the impact of an economic activity on outsiders. For example, the market may ignore the costs imposed on outsiders by a firm polluting the environment.
- *Public goods*, such as national defence. How much defence would be provided if it were left to the market?
- Where there is incomplete or *asymmetric information* or uncertainty.

Abuse of market power is best tackled through *antitrust* policy. Externalities can be reduced through *regulation*, a tax or subsidy, or by using property rights to force the market to take into account the *welfare* of all who are affected by an economic activity. The *supply* of public goods can be ensured by compelling everybody to pay for them through the tax system.'

Source: *The Economist* (2009).

Note: *Antitrust policy* is the North American expression which refers to public policy relating to monopoly and competition. The basic explanation of *asymmetric information* is that it occurs when one or more parties to an economic transaction has more or better relevant information than other parties – putting them at an advantage if the information is not shared.

---

1 One approach is to use subsidies and taxes to modify market prices as has been reviewed above (in Box 8.2), modifying the 'real' income distribution in the process.

2 A second approach is for the government to become the direct provider of particular goods and services if there is a fundamental failure by the market to deliver essentials.

3 A third approach is for the government to adopt sets of regulations which modify market behaviour – such as regulations relating to competition, to the environment and public health, and to the labour market. The application of regulations, which are widespread in developed industrial economies, involves transactions costs and systems of penalties which can be the subject of detailed economic analysis.

4 A fourth approach is to impose certain market characteristics on the supply of goods and services where no market has previously existed, such as through the requirement of competitive tenders and bids within a contracting process – sometimes referred to as the Principal:Agency approach. This has been widely adopted by governments in recent years as one variant of 'privatisation', and is also referred to as a 'contestable markets' approach.

Much of this approach to government intervention involves 'arm's length' management consistent with conventional economic 'wisdom'. A major problem for developing countries is that the institutional framework and capacity is often insufficiently robust to fully support these approaches, which have been transferred from experience in developed market economies.

One aspect of market failure relates to what are often referred to as 'Adverse Selection' and as 'Moral Hazard'. These expressions are often used with an assumption that everybody understands what the writer (or speaker) means – however, this often conceals the fact that the writer (or speaker) is far from clear about what they themselves mean. While both terms are used with equal relevance and validity in developing and in developed industrial countries, they perhaps have greater significance in developing economies where legal redress for the effects of misrepresentation and fraud is considerably more difficult to achieve. *The Economist* website's A to Z of economics refers to both Adverse Selection and Moral Hazard principally in the context of insurance markets (2009), but this is undesirably restrictive. Essentially 'Adverse Selection' refers to a combination of circumstances through which decisions (such as awarding a contract or making a decision to purchase) are based on falsified or inadequate information – so that if the correct or full information had been available, a different decision would have been made. 'Moral Hazard' refers to a situation where, a decision having been made and implemented (for example, the award of a contract), the contracted party does not abide by the terms of the contract. An example of this would be where a loan has been obtained, but the borrower uses the funds for a different purpose to that intended by the lender. Another example would be where a builder obtains a contract (and payment) for the undertaking of work which is not completed satisfactorily (if it is even started). A third example would be where a loan has been received by a borrower who has no intention of completing repayment. In the context of good governance, 'Adverse Selection' and 'Moral Hazard' represent unacceptable features of markets. Again, the supposition is that these issues are more of a problem in developing countries by comparison with developed industrial countries.

One issue which excites considerable controversy within the economics profession relates to the determination of an appropriate rate of discount (or rate of time preference) to be used in environmental analysis, and particularly in the analysis of climate change. The rate of discount or of time preference concerns the view (by an individual, by an institution, or by society as a whole) of the relative value of economic resources a) at the present time and b) in the future, and it is usually expressed as a percentage (akin to a rate of interest). With the long time horizons which apply to many environmental issues (such as climate change in general or the significance of ozone layer depletion in particular), there is considerable agreement amongst economists that a low rate of discount should be used – in the region of perhaps 2.5 per cent, while the institutional view of bodies such as the World Bank has been that a universal discount rate of 10 per cent to 12 per cent should be used across the board for all economic analysis of future investments/ events.

In the UK, the Department for International Development has usually tended to follow the advice of the World Bank on the appropriate discount rate to use for the analysis of investments and/or policy options in developing countries. This means that a discount rate of 10 per cent to 12 per cent (in real terms) would be used. On the other hand the UK Treasury currently uses a discount rate of about 3.5 per cent for the analysis of public sector investments and/or policy options (HM Treasury, 2008). In his overview of the economics of climate change, Stern justifies using a discount rate for the very long-term perspective which is considerably lower than those which have conventionally been adopted for the analysis of development projects and policy (Stern, 2006: Chapter 2), and lower than the Treasury's rate. The quite wide range of discount rates described here is further evidence of a certain lack of consensus within the economics profession.

Finally in this section two important issues, one theoretical and the other empirical, have been included in Boxes 8.4 and 8.5. One is the 'theory of second best', which gives a methodological basis for government intervention in the context of market failure. The other is 'cash transfers', which have become an integral part of development policy in recent years.

Box 8.4

The 'theory of second best'

The general theory of second best was initially expressed by Lipsey and Lancaster (1956). The basic proposition starts from the fact that the entire set of conditions or assumptions of perfect competition or of the perfect market are embodied in what are known as 'Pareto conditions'. These conditions are not intended to describe reality, are never found in practice and exist as the basis for theoretical analysis. The implication of this is that conclusions drawn from models based on these neo-classical Paretian conditions (relating to what is termed 'Pareto optimality') are unlikely to be logically appropriate for policy guidance. In particular, if one of the crucial conditions or assumptions of the theory does not apply (e.g. small numbers of sellers or of buyers rather than large numbers), then any conclusions from the model do not necessarily follow logically. In these circumstances, the 'second best' solution – a form of government intervention – can be preferred based on intelligent analysis of the 'imperfect' situation.

A particularly important property of the Paretian system is that economic judgements about improvements in economic welfare can only be made if, as the result of a policy change, at least one person is better off and nobody is worse off. Since, in most cases, policy implementation results in some people being better off and others worse off, it is not possible to make any judgement about whether overall welfare has improved on purely economic grounds. However, it is possible to make such judgements on non-economic grounds, involving value judgements about the relative significance of the welfare of those who are better off and those who are worse off. There are economic concepts which can help to make such value judgements systematic, but ultimately they are political rather than economic judgements.

In practice, it can be said that the theory of the second best gives the intellectual justification for government intervention in conditions of 'market failure'. Conditions of 'government failure' are likely to arise where the analysis on which policy has been based is inadequate or where no such policy analysis has been undertaken. In many cases, the interaction of special interest groups (including corruption) can lead to policy implementation deviating from a logic based on careful analysis – which is a major reason why neo-liberals (and even some more 'heterodox' economists) are wary of government intervention.

Box 8.5

'Cash transfers'

'Cash transfers' (often 'conditional' or 'targeted') have been introduced to development programmes as a means of quickly dispersing funds which have short-term poverty reduction (social protection) or disaster relief objectives. Two of the main propositions made by supporters of cash transfers are that a) transactions costs for transfers in 'kind' can be very high, b) poor people often have a better view of expenditure priorities for the achievement of welfare improvements (poverty reduction) than bureaucrats (and economists), and c) transfers with short-term objectives do not clash with economists' efficiency criteria (Ravallion, 2003).

The economic argument against cash transfers for short-term poverty reduction has been partly based on the fact that 'redistribution', through increasing consumption, conflicts with investment in longer-term growth and development arising from investment in productive activity. However, there is increasing evidence that high levels of income inequality actually detract from growth and development objectives, so that higher 'social protection' transfers can be 'efficient' in economic terms through leading to greater equality. However, it is also necessary to allow for the fact that priorities based at a 'grass-roots' level may relate to short-term time horizons, and could lead to lower expenditure on development programmes such as in education and health (which have a significant long-term poverty reduction role) than is desirable. For this reason, 'conditional' transfers have been based on requirements such as that children attend school and visit health centres (Rawlings and Rubio, 2005).

Another economic argument against cash transfers is that 'grass-roots' priorities may not take sufficient account of external costs and benefits and of complementarities. For example, higher education may be given a lower priority relative to primary education than is desirable in long-run terms, and external benefits associated with preventive medicine, or collective water and sanitation programmes may be given a low priority (relative to curative medicine).

The UNDP International Poverty Centre (www.undp-povertycentre.org) has produced a number of studies of cash transfer systems, as has the Overseas Development Institute (www.odi.org.uk). The Chronic Poverty Centre also has a significant number of publications on this issue (www. chronicpoverty.org).

8.4 Neo-liberalism and the Washington Consensus

The Washington Consensus has been regarded as representing the essence of neo-liberalism. One of the major tenets of neo-liberalism is promotion of a 'free market', taking the position that the unfettered operation of markets is likely to lead to the types of 'efficient' outcomes arising from the economist's model 'perfect market'. 'Liberalisation' of markets, providing the conditions for freedom of entry and exit for example, is therefore central to the neo-liberal creed. Other central features of the neo-liberal position are that in order to achieve freedom of entry and exit, responsiveness to market conditions and competition, it is essential that most economic activity takes place within the private sector, with as little government intervention as possible, including low or zero subsidies, with little regulation, and a small government 'footprint' on the economy (which includes as low an overall level of taxation as is possible). However, even neo-liberals would recognise that policing, national security and defence and a number of other activities have to be largely provided by governments, and require funding from public taxation.

It takes only a small leap of imagination to link the economist's theoretical concept of the 'perfect market' to the neo-liberals' concept of the 'free market'. What the economist describes as 'imperfections' of markets are actually very common occurrences (refer to Box 7.2 in Chapter 7). For example, small numbers of sellers and buyers who have asymmetric market power with restricted entry and exit is a 'usual' situation for many markets, and the policy issue is more that of what form of intervention is appropriate rather than to advocate no intervention. Entry to markets by new suppliers can be constrained by a) the amount of capital investment needed to join market suppliers, b) restricted access to specialist technical knowledge, c) restricted access to markets due to specialist information and pre-existing contractual arrangements, and d) restricted access to supplies of essential inputs.

There is another play on words which involves the juxtaposition of the terms 'neo-classical economics' and 'neo-liberal'. To a significant extent, neo-classical microeconomics has its basis in 'marginal' analysis based on the assumptions of the 'perfect market' (refer to Box 7.2 in Chapter 7). Neo-classical economic analysis aspires to a high degree of objectivity and scientific method. Neo-liberalism aspires to neither objectivity nor to scientific method, largely because it essentially represents an ideological position advocating the free play of market forces despite asymmetries of market power. Those who are critical of neo-liberalism take the view that the 'free play of market forces' reinforces and extends existing economic imbalances and inequalities. It is in this context that the Washington Consensus needs to be examined.

Table 8.1 summarises the main features of the Washington Consensus. The objective of the table, and in the brief discussion which follows, is to distinguish between those aspects of the Washington Consensus a) which contain a widely acceptable diagnosis and prescription providing a basis for better and more equitable economic performance, and b) which critics regard as having a contestable economic and ideological basis.

The Washington Consensus, as practised by the international financial institutions, is widely regarded by heterodox economists as an explicit programme for the dismantling of the 'developmental state' and its replacement with a more market-friendly approach based on neo-liberal principles (see, for example, Toye, 2003 and Fine, 2006b). The advocates of the Washington Consensus regarded the developmental state as being riddled with distortions and dysfunctional policies associated with government intervention – government failure rather than market failure. The 'key anomalies' were perceived to include markets constrained by government and public sector action, and the principal policy vehicle which was to address this 'problem' was to be liberalisation in all its forms (as outlined in Table 8.1). The enthusiasts for the Washington Consensus stimulated economists who were uneasy about the neo-liberal agenda, but who nevertheless had misgivings about the 'developmental state', to re-think the 'role of the government' in economic development. Two clearly written examples of this approach are papers by Stiglitz (1997) and Adelman (2001).

The practice of the Washington Consensus was embodied in Structural Adjustment Programmes and Economic Reform applied to developing countries from the late 1970s until the mid-1990s. The Washington Consensus was formally articulated in the early 1990s as an *ex post* description of the overall approach to Adjustment and Reform (Williamson, 1990, 1993, 1994, 2000, 2004). Criticism of the Washington Consensus and of Structural Adjustment Programmes

Table 8.1 The Washington Consensus

Main issue	Characterisation of the problem	Content of the Washington Consensus	Implications as practised by the IFIs	Outcomes expected
Distortions in post-colonial economies have been caused by state intervention in the form of dysfunctional policies.	1 Significant government deficits and high rates of inflation.	1 Fiscal discipline.	Universal economic principles – one size fits all:	1 Macroeconomic stability.
	2 Mismanagement of public expenditure.	2 Prioritising public expenditure.	Focus on:	2 More predictable and manageable economy.
	3 Inadequate tax collection.	3 Tax reform.	1 Short–term allocative efficiency rather than on growth, equity and poverty reduction.	3 Firmer basis for achievement of economic and social policy objectives.
	4 Negative real interest rates and inadequate financial sector management.	4 Financial liberalisation.	2 Macroeconomic stability.	4 The end of the 'developmental state'.
	5 Overvalued foreign exchange rates with disincentive for exports.	5 Market determined exchange rates.	3 Open economies – trade and financial liberalisation vis-à-vis the rest of the world regardless of the impact on the domestic economy.	
	6 Trade policy provides random and dysfunctional protection for domestic production.	6 Trade liberalisation.	4 Liberalisation and de-regulation of domestic markets.	
	7 Economic policy environment discourages foreign direct investment.	7 Openness to foreign direct investment.		

Table 8.1 (continued)

Main issue	Characterisation of the problem	Content of the Washington Consensus	Implications as practised by the IFIs	Outcomes expected
	8 Many public institutions operate with significant deficits and inadequate replacement and new investment.	8 Privatisation.		
	9 Many regulations provide disincentives for investment and innovation.	9 Regulatory reform and deregulation.		
	10 Inadequate protection of personal and property rights.	10 Property rights reform.		

Sources: based on Williamson (1993, 1994, 2004).

has been robust and widespread. One of the leading critics has been Stiglitz (1989, 1998a, 1998b, 2002, 2004) and a valuable overview of the evolution of this controversial set of issues has been provided by Gore (2000). Some critics, including Stiglitz, have referred to a post-Washington Consensus, which represents a 'synthesis' of elements of both the Washington Consensus and the 'developmental state'. However Rodrik (2002: 1), a leading academic critic of liberalisation, has labelled the post-Washington Consensus as 'unfeasible, inappropriate and irrelevant' and dubbed it the 'augmented-Washington Consensus', arguing that it attempted to describe desirable features of 'development' but not how to attain them, and added ten policy areas to Williamson's original list to form this implicit 'augmented WC'. This 'augmented' version of the WC (the post-Washington Consensus) added corporate governance, anti-corruption, flexible labour markets, WTO agreements, financial codes and standards, 'prudent' capital account liberalisation, non-intermediate exchange rate regimes, an independent central bank and inflation targeting, social safety nets, and targeted poverty reduction to the original ten canons of policy. Some of Rodrik's arguments relating to the WC have been elaborated in his book *One Economics: Many Recipes* which includes an emphasis on the need to tailor policy to local circumstances rather than to broad universalistic prescription (2007).

While the Washington Consensus articulated a set of domestic policies, many of the problems of poorer developing countries related to high levels of international indebtedness which had been built up during earlier periods of international economic instability and domestic economic mismanagement (Todaro and Smith, 2008: Chapter 13). It was not until the adoption of the Heavily Indebted Poor Countries (HIPC) initiative in the late 1990s that these problems were seriously addressed, to which the Multilateral Debt Relief Initiative (MDRI) was added in 2005 (UNCTAD, 2008: 38–41; World Bank, 2009).

Maxwell (2005: 1–2) argues that the post-Washington Consensus is still 'incomplete' and he tentatively proposed a new 'meta-narrative': 'the WC has been replaced by a new and improved orthodoxy, called here the "meta-narrative". It emphasises the Millennium Development Goals (MDGs) as an over-arching framework, and lays out the link between the MDGs, nationally owned poverty reduction strategies, macro-economic policy (including trade), effective public expenditure management, and harmonised aid in support of good governance and good policies' (Maxwell, 2005: v). In this context, a 'meta-narrative' is the 'big picture' or – in other words – the overall view of the world within which discussion takes place.

8.5 International aid

The role of international aid in the development process has been one of the most controversial issues in the study of developing countries over the last few decades. Most development economists tend to be 'in favour' of international aid, with their reasons relating to its contribution to economic growth and to poverty reduction in particular. The contributions of aid to economic growth are regarded as being both 'direct', such as through funding investment in social and economic infrastructure and providing technical assistance (direct personnel as well as the funding of scholarships and training), and 'indirect', such as through contributing to a better policy environment. A good recent collection of papers on aid and development was edited by Tarp and Hjertholm (2000), and a comprehensive review of aid was undertaken by Riddell (2007).

However, there are also schools of thought which are dismissive of aid. The negative view arises at both ends of an ideological spectrum – the 'free-marketeers' who regard aid as an undesirable form of government intervention (i.e. related to both 'donor' and 'recipient' governments), and Marxists and neo-Marxists who regard aid as an extension of neo-imperialist socio-economic structures. The discussion in the remainder of this section will focus on two questions – first, how 'effective' is aid and, second, how can the effectiveness of aid be improved?

Whether aid is effective or not depends initially on which definition of effectiveness is used. Journalistic and populist approaches to this question tend to focus on the detail of aid projects and programmes – do particular water systems, or schools and hospitals, work as intended (DFID, 2006)? However, economists are concerned with broader and more systematic pictures which include, for example, whether aid increases the level of investment in recipient countries, whether aid increases the rate of economic growth, and whether aid has a significant effect on poverty reduction (see, for example, Nixson, 2007–8; McGillivray et al., 2006; Morrissey, 2001). While the economists' questions are clear and very relevant, the answers require detailed analysis based on a rigorous methodology.

One of the key controversies for development economists concerns the relationship between aid, the policy environment and economic growth, and this is shown in Figure 8.1. The first segment of the figure (A) shows a direct and simple relationship between aid, investment and growth. The second (B) shows that aid may improve the policy environment, which in turn may increase investment (attracting both domestic and foreign investment) and lead to higher economic growth. The third (C) shows that a better policy environment may lead to higher levels of aid, investment and growth. The essential question is – what is the causal relationship between these variables, and in which direction does the causation work? Quantitative analysis (usually using forms of regression) which does not take account of this complex web of potential causation is likely to be of little interest or value from an economic policy standpoint.

Burnside and Dollar (2000), both from the World Bank and representing essentially a mainstream 'neo-classical' approach, published an influential article which proved to be significant but controversial. Their analysis found that aid increased the rate of economic growth, but that this effect was dependent upon the 'policy environment'. Methodologically this finding raised the question of whether it was the aid or the policy environment which accounted for the higher economic growth rate. It is conceivable that the better growth performance was due more to the better policy environment (better economic management or better governance) than it was to

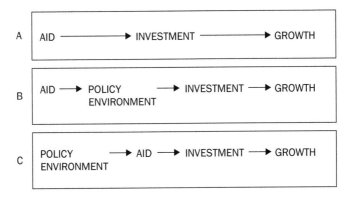

8.1 The relationship between aid and economic growth

the international aid flows. Statistical analysis is not really capable of answering this question. Logically this line of questioning raised yet another issue: that of how it is possible to compare the quality of the policy environment both over time within one country and at any one time between countries (this is known as time series and cross-country analysis respectively). The World Bank has attempted to respond to this issue through the development of a Country Policy and Institutional Assessment (CPIA) (World Bank, 2007c) which is explained and discussed in Box 8.6. Because the CPIA provides a quantified measure which represents the quality of governance, this measure can be included directly in quantitative analysis of the relationship between aid levels, the policy environment and economic performance. However, the derivation of the CPIA can be questioned on methodological grounds so that it represents a 'work in progress' measure of the policy environment.

The most controversial aspect of Burnside and Dollar's argument was the suggestion that governance (the policy environment) should be a criterion for the allocation of aid between countries by 'donors'. This would imply that countries with 'poor' governance would receive relatively less aid per capita, and countries with 'good' governance would receive relatively more aid per capita – other things being equal. Amongst other objections to this proposition was the obvious one – that poor countries (i.e. low income per capita and implicitly high levels of poverty) which would be regarded as requiring the highest levels of aid were also the most likely to have difficulty in achieving a 'high' CPIA measure. Equally, countries emerging from a period of conflict, and 'deserving' of relatively high levels of aid on other criteria, would fall foul of a CPIA-oriented criterion of aid allocation. Nissanke (2007) and Nissanke and Ferrarini (2007) are particularly incisive in criticism of this approach, and where it has been adopted by the international aid institutions it is only one of a range of criteria used for the allocation of international aid. It can be noted that the procedures relating to aid allocation and management are referred to as the 'aid architecture' while the significant 'policy' issue is still that of how the effectiveness of aid can be improved.

Box 8.6

The Country Policy and Institutional Assessment (CPIA)

The objective of the CPIA has been to create a numerical value for the phenomenon of 'governance' or 'policy environment'. For economists, the existence of a numerical value of this type for a 'qualitative' phenomenon means that it can be included in quantitative analysis – for example, comparing the policy environment with various aspects of economic performance.

The CPIA has changed several times since it was first developed, but is now based on sixteen criteria consisting of an Economic Management cluster (Macroeconomic management, Fiscal Policy and Debt Policy), a Structural cluster (Trade, Financial Sector and the Business Regulatory Environment), a Policies for Social Inclusion and Equity cluster (Gender Equality, Equity of Public Resource Use, Building Human Resources, Social Protection and Labour, Policies and Institutions for Environmental Sustainability) and a Public Sector Management and Institutions cluster (Property Rights and Rule-based Governance, Quality of Budgetary and Financial Management, Efficiency of Revenue Mobilisation, Quality of Public Administration, Transparency Accountability and Corruption in the Public Sector).

Each of the sixteen variables are assigned a mark between 1 (low performance) and 6 (high performance) and the average scores are arrived at by cluster and for overall 'governance performance'.

There are several methodological problems with this CPIA measure. The respective significance of each of the individual sixteen criteria and four clusters (i.e. the relative weights adopted) are open to question. The people responsible for making the assessments within any particular country need to maintain consistency between criteria and between years. Comparisons between countries are only possible if it is possible to be confident that the assessment criteria have been implemented uniformly. Also, it is clear that the ranking within each criterion is based on ordinal principles while adding together the scores for each criterion and dividing by the number of criteria uses cardinal mathematical principles – a mixing of ordinal and cardinal principles which is questionable.

However, the CPIA is the best approach which we have for the moment, and the scores arrived at are interesting but they need to be treated with caution and respect.

Source: World Bank (2007c).

8.6 The environment and development policy

Elsewhere in this chapter, and in other parts of this book, there have been some references to the economic dimensions of environmental issues in international development. For example, section 8.3 includes a brief discussion concerning the appropriate rate of discount to use in the economic analysis of long-term environmental phenomena. Section 8.3 also reviews the role of taxes and subsidies (including transactions costs) as policy instruments which can be used with varying degrees of success in achieving policy objectives, including those relating to environmental policy. In this section, we aim to discuss a few more significant environmental issues which relate to developing countries from an economic perspective.

Climate change is the biggest environmental challenge facing the planet, with incontrovertible evidence that recent and projected human activity (anthropocene) threatens to significantly raise global temperatures with serious impacts on agriculture, water supply, sea levels, disease and living conditions (Stern, 2006). Action to address this challenge will involve considerable expenditure on investment in new technologies which reduce and reverse the causes of temperature increases (UNFCCC, 2008). The priorities for this investment will have to be based on careful economic analysis but, more significantly for the focus of this book, it is necessary to consider the respective positions of higher-income developed economies and the lower-income developing countries which are our main concern.

The human activity which has been responsible for almost all of the recent rise in global temperatures has taken place in higher-income developed countries, but the environmental and economic impact of the temperature rise has affected lower-income developing countries at least as much as the higher-income countries. In economic terminology, this impact on the lower-income countries is a classic example of an external diseconomy and of market failure. The lower-income countries experience economic costs (or reduced benefits which amount to the same thing) which are not caused by themselves but international markets do not provide any form of compensation from the countries which have caused the costs or sacrificed benefits.

Discussions taking place in preparation for the 2009 Copenhagen meeting on global climate change included the issue of how the higher-income countries could provide a degree of compensation (not through 'markets' but through government contributions) to lower-income countries for the impact of climate change, and how much compensation might be provided (UNFCCC, 2008). This is an example of how an economic concept is the basis for substantial financial commitments in the international development context.

Another important consideration for developing countries is that of the prioritisation of environmental issues. There are two main points which can be made here. First, that the highest priority will usually be given to environmental impacts which have the most serious economic implications. By economic implications we mean either that there is a need to commit significant amounts of expenditure to 'correcting' for environmental impact, or that environmental impact imposes significant income loss (through, for example, loss of production). Second, it is necessary to distinguish between environmental impacts which are relatively easily (i.e. without major expenditure) reversible, and those which are impossible to reverse or which are reversible only with very high levels of expenditure.

A number of important conclusions can be drawn from this approach. First, the environmental issues which are of most concern to developing countries (apart from global climate change) are very likely to be different from those which most concern developed industrial countries. For example, soil degradation, deforestation, water conservation and fuelwood will be high on the list of priorities for most sub-Saharan African countries. On the other hand, for highly urbanised developed industrial countries the highest priorities will probably include energy conservation (relating to transport and to industrial/domestic demands), waste disposal and congested transport systems.

Statistical evidence relating to this issue of environmental priorities is difficult to produce, and is difficult to find. However, some is available for Ghana, where there has been an encouraging amount of work on this issue. Environmental degradation arising from agriculture comes from annual crops such as vegetables, grains and tubers. Tree crops, such as cocoa, coffee and rubber have some environmental benefits because of their vegetative cover. However, the primary process of converting forests into tree crops is a first step degradation. The farming practice of slash and burn exposes the soil to the full intensity of wind and rain leading to soil erosion. The consequence is soil and nutrient loss and finally productivity loss. Soil erosion also causes siltation of facilities such as dams, lakes and streams. The total estimated gross costs of environmental degradation in Ghana was 41,305,000 cedis (US$127 million) in 1989. Soil erosion and other forms of land degradation accounted for 70 per cent of the gross environmental damage costs for that year (Convery and Tutu, 1991; Convery, 1995: Chapter 8).

Another dimension of the contrast between high-income developed industrial countries and low-income developing countries is that of the appropriate set of policy instruments which can be adopted in addressing environmental policy priorities. It should be clear that, if environmental priorities differ between these two (admittedly somewhat stylised) groups of countries, then the mix of policies and policy instruments which will be adopted will differ. Policy measures associated with energy conservation in developed countries will not be relevant to soil conservation in developing countries. However, the differences between the socio-economic characteristics of the two groups of countries will imply that even for the same type of environmental 'problem' different types of policy instrument might be appropriate. For example, in higher-income developed countries part of the current problem associated with waste disposal is to find a way of shifting away from putting waste into 'landfill sites' and towards higher levels of re-cycling. This has involved the introduction of taxes (financial penalties) on the use of landfill sites, and the development of re-cycling systems (and incineration) for waste disposal. There is also pressure for a reduction in, and changes to, the packaging technology for consumer goods. In many developing countries, the level and patterns of consumption have not been associated with quite as much domestic waste as has been experienced in higher-income countries, and the types of waste are likely to differ. Developing countries do not have the same extent of prepared, pre-cooked and frozen food as that found in more developed countries, and containers (such as glass and plastic bottles) tend to be re-used spontaneously within established markets rather than to be re-cycled as a type of industrial input (Tribe, 1996).

The final issue in this brief environmental section relates to the economically 'optimum' level of pollution abatement – which can also be viewed as determining criteria for an optimum level of pollution. Although this may appear, at first sight, to be a somewhat bizarre concept, it is actually directly relevant in a conceptual context to the setting of regulations about acceptable (and unacceptable) levels of pollution. Complete lack of pollution (i.e. 100 per cent abatement) would be very expensive to achieve, and for different types of pollution lower levels are usually tolerable and can be achieved at considerably lower cost. The issue which we need to address here is that of how Figure 8.2 is relevant to developing countries.

Although Figure 8.2 represents a conceptual framework, it is usually very difficult to quantify the variables involved. What is clear is that the values for marginal benefits and marginal costs arising from pollution abatement (shown by the diagonal lines) will be determined by a combination of physical values (quantities) and economic values (prices). The economic values will vary from one location (or country) to another depending upon local circumstances. The physical values will also vary between locations depending upon the nature of the pollution and the balance between natural and 'artificial' dissipation of the pollution. One of the main implications of this is that what would be an optimum level of pollution abatement in one location is not necessarily applicable in other locations – so that regulations designed for one location are not necessarily relevant for other locations.

An example might make this issue clearer. Breweries tend to emit significant amounts of solid waste matter suspended in water, unless their emissions are cleaned before they are disposed of. Most of these emissions are bio-degradable, but this process takes time – the amount

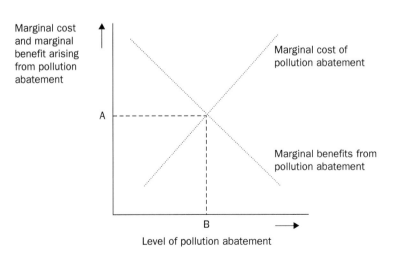

8.2 *The optimum level of pollution abatement*

of time depending partly upon ecological conditions. If the ambient temperature is relatively high, the natural process of degrading the solid waste matter may be acceptable without recourse to expensive technology provided that measures are taken to keep the waste matter separate from areas of water which are 'sensitive' during the degrading process. The lower the ambient temperature, the more likely it is that cleansing technologies will be necessary in order to achieve acceptable degrees of cleanliness of emissions. Although it has not been possible to provide scientific 'chapter and verse' for this example, it should be clear that the principles which are being invoked are directly relevant to the issue of variability of appropriate environmental regulations on pollution levels. In the context of the brewery emissions example, this implies that regulations designed for application in countries with a temperate climate are not directly transferable to countries with a tropical climate, such as from Western European countries to sub-Saharan African countries.

8.7 Summary

- In the context of development policy, and of economic policy, 'conditionality' has been a major issue in recent decades, relating to the application of 'policy conditions' to international assistance.
- A distinction can be made between macroeconomic and microeconomic policy.
- There is controversy over the extent of neo-liberal influence over policy conditionality – with a distinction between technical and ideological aspects of 'conditions'.
- Macroeconomic management includes balances between consumption and savings, public and private sectors, credit creation and balanced public sector budgets, and management of international payments and the foreign exchange rate.
- Microeconomic management includes policy instruments such as subsidies, and policy areas such as the pricing of public services and the financial sustainability of institutions.
- A major economic policy issue is that of the extent to which governments should take action to redress 'market failures' – which include the phenomena of 'adverse selection' and 'moral hazard'.
- Within the context of conditionality and guidelines for policy formulation, the Washington Consensus has been a major controversial feature, and some writers have referred to a 'post-Washington Consensus'.

- The contribution of international aid to sustained socio-economic development has been another controversial area. Two particular issues relate to the effectiveness of aid and to the criteria for the allocation of aid between developing countries (including the issue of 'aid architecture' which refers to management procedures within aid institutions).
- Policy towards the environment needs contrasting approaches in developing countries and developed industrial countries, with different priorities and mixes of policy instruments being relevant in distinct socio-economic and ecological contexts.

Questions for discussion

1 What is the distinction between 'microeconomic' and 'macroeconomic' policy in the development context?

2 What were the main objectives of the 'Structural Adjustment Programmes' of the 1980s and 1990s and to what extent were they achieved?

3 What do you understand by the 'Washington Consensus' and how has it been linked to 'Structural Adjustment' and to 'Economic Reform'?

4 What are the main economic policy instruments which can be used for the achievement of microeconomic development objectives?

5 What is 'market failure'?

6 To what extent can 'international aid' help with the achievement of development objectives?

7 What are the main environmental policy issues in developing countries?

Suggested further reading

Gore, C. 2000. The Rise and Fall of the Washington Consensus as Paradigm for Developing Countries. *World Development*. 28(5): 789–804.

Killick, T., Gunatilaka, R. and Marr, R. 1998. *Aid and the Political Economy of Policy Change*. London: Routledge.

Stiglitz, J. 1997. The Role of Government in Economic Development.

Proceedings of the Annual World Bank Conference on Development Economics 1996. Washington: The World Bank, 11–26.

Todaro, M. and Smith, S. 2008. *Economic Development* (10th edn). London: Pearson Addison Wesley, Chapters 10 and 16.

Toye, J. F. J. 2003. Changing Perspectives in Development Economics. In Chang, H.-J. (ed.) *Rethinking Development Economics*. London: Anthem Press, 21–40.

Economic concepts used in this chapter

1 International aid process
   Policy conditionality
   Technical assistance

2 'Political economy'
   Neo-liberalism
   Washington Consensus
   Free market
   Structuralist/heterodox economists
   Adverse Selection/Moral Hazard
   Structural Adjustment Programmes/Economic Reform
   Market friendly

3 Economic analysis
   Neo-classical economics
   Mainstream neo-classical economics
   Macroeconomics
   Microeconomics
   Economic efficiency
   Macroeconomic projection model
   Working capital
   'Pay-back period'
   Inflation
   Directly productive activities
   Infrastructure
   Marginal analysis
   Perfect market
   Supply price
   Market clearing
   Demand and supply elasticities
   Transactions costs
   Household expenditure
   Transfer payments

Capital gain
Open market prices
Incremental approach/with-without analysis
Externalities/external economies and diseconomies
Public goods
Asymmetric information
Rate of discount/rate of time preference
Theory of second best
Time-series and cross-section analysis
Economic policy
Market liberalisation
Subsidies
Full cost recovery
Subventions
Financial discipline
Policy instruments
Equity
Reciprocity
Anti-trust policy
Contestable markets
Principal:Agency approach
Cash transfers

4 Public finance and expenditure analysis
Public Investment Programme
Medium Term Expenditure Framework
Public investment expenditure
Recurrent expenditure
Domestic borrowing
Accelerator effect
Economic governance
Government deficit
Credit creation/'printing of money'
'Crowding out' and 'crowding in'
Fiscal neutrality

5 International economic policy
International contagion
Dutch Disease
Exchange rate appreciation
Competitiveness of exports

9 Poverty, inequality and development economists

9.1 Introduction

This chapter is concerned with the contribution of development economics to the analysis of poverty and inequality. The discussion relates both to conceptual contributions as well as contributions to policy/strategy and to indicators of poverty and inequality. It is argued that contemporary development economists have been at the forefront in the evolution of thinking on poverty and inequality. Indeed, in many senses development economists have led development studies away from approaches that are primarily economic while still emphasising the significance of the economic dimension of poverty analysis.

In recent years, development studies and even development economics have moved away from purely economic approaches to poverty analysis, both in goals and in policy. There has been a shift, to varying degrees, from GDP per capita (see Box 5.1 in Chapter 5) and income poverty to multi-dimensional poverty exemplified in the UN poverty targets for 2015, otherwise known as the Millennium Development Goals (MDGs), and more recently the concept of 'wellbeing'. Emphasis on growth maximisation has been replaced by Poverty Reduction Strategy Papers (PRSPs) with their over-riding concern for poverty and inequality within the context of good economic management.

On two central development studies issues in particular – poverty and inequality – development economics has made some central contributions. Indeed, Kanbur and Squire (2001: 183) have placed poverty reduction 'at the heart' of development economics. Such interest is not new in economics either. It drove not only the 'founding fathers' of quantitative economics, such as Petty and Quesnay, but also the 'pioneers' of political economy – Marx, Smith, Ricardo, Malthus and Mill (Anand and Sen, 2000: 2031). Since 1945, a succession of leading development economists have made very significant contributions to the literature relating to poverty and/or inequality including Kanbur, Myrdal, Ravallion, Seers, Sen and Streeten – as is discussed later in this chapter.

In this chapter, the criteria for choosing the 'contributions' are that they represent 'major' shifts in thinking and are concepts or approaches that have become commonly used in development research and policy. We are in no doubt that this selection process is subjective and that there are many more contributions which could have been added, but space constraints demand limits. Section 9.2 discusses selected contributions of development economics to the analysis of poverty. Section 9.3 addresses selected contributions of development economics to the analysis of inequality. Section 9.4 discusses indicators of economic and non-economic poverty and inequality. Section 9.5 concludes.

9.2 Selected contributions by development economists to the analysis of poverty

The meaning of poverty

Development economics has made a major contribution to debates on the meaning and measurement of poverty and has led discussion away from purely economic approaches to recognising multi-disciplinary understandings of poverty. Box 9.1 outlines the concepts of 'absolute' and 'relative' poverty, which are fundamental to much of the discussion in this chapter.

In the 1970s, Dudley Seers – an economist – started what might appear as an almost seamless paradigm shift from an economic definition of development, largely measured in terms of GDP per capita, to a broader definition extending beyond a prime concern with economics to one more concerned with poverty and inequality. Seers (1969: 3–4) famously asked:

> What has been happening to poverty? What has been happening to
> unemployment? What has been happening to inequality? . . . If one
> or two of these central problems have been growing worse,
> especially if all three have, it would be strange to call the result
> 'development', even if per capita income has soared.

Further major contributions on 'basic needs' were made by other
development economists, notably Paul Streeten (see Hicks and
Streeten, 1979; Streeten, 1984) and economists at the ILO (1976,
1977). At the same time, dissatisfaction with GDP per capita
stimulated work on 'levels of living' indicators – again by
development economists (see, for example, Baster, 1972, 1979;
McGranahan et al., 1985; Morris, 1979; UNRISD, 1970). Although
there was a lull in interest in the 1980s due to attention to the debt
crisis and adjustment (neo-liberals were more concerned with
stabilisation and growth and less concerned with poverty at that time),
concerns that 'basic needs' could not be guaranteed by GDP growth,
or even by raising income amongst the poorest, were voiced again as a
result of the social impacts of adjustment (Kanbur and Lustig, 1999).

In the 1990s, development thinking and policy were fundamentally
reshaped by the work of Amartya Sen – himself a development
economist – and the new annual report, launched in 1990 by the
United Nations Development Programme (UNDP), the *Human
Development Report* (various years). Sen developed the 'Capability
Approach' (CA) or 'Human Development' and since then the *Human
Development Reports* have played a major role in repositioning
poverty from the periphery to the core of attention (Sagar and Najam,
1999: 743). By 2000, the World Bank's *World Development Report*
was quoting Sen (1999) on its opening page (World Bank, 2000: 15).

Sen's major contribution was a critique of the then current concept of
'wellbeing' which had been defined within economics as 'desire
fulfilment' (Bentham's psychic utility) or consumption. In particular,
wellbeing was measured by the proxy of income – GDP per capita –
and Sen argued that this did not take sufficient evaluative account of
the physical condition of the individual and of a person's 'capabilities'
(Sen, 1979; Sen, 1993: 31). In the Capability Approach, the goal of
development is not maximisation of GDP per capita but 'a process of
expanding the real freedoms that people enjoy': that is to say that
poverty is 'capability deprivation' and poverty reduction is a question
of expanding capabilities (Sen, 1999: 1, 18). According to Sen (1988)
freedom has two aspects. An 'opportunity' freedom refers to the

Box 9.1

The concepts of absolute and relative poverty

The concepts of absolute and relative poverty are generally associated with economic- or income-based definitions of poverty (although conceptually they could be associated with non-economic definitions of poverty such as 'social exclusion' – see Box 9.3).

For absolute poverty we can specify a 'poverty line', with those below the line being regarded as in absolute poverty, and those above the line not in absolute poverty. The percentage of any given population below the poverty line is the 'headcount index' of (absolute) poverty, and comparisons are widely made between the headcount index of poverty for different countries at particular points of time (cross-country or cross-section comparisons) and between different points of time for the same country (historical or time-series comparisons).

Some definitions of 'absolute poverty' are based on a poverty line represented by the estimated physiological requirements for the maintenance of wellbeing (such as a 2,100 calorie level). These physiological requirements can then be valued in order to arrive at an associated level of expenditure. Clearly this type of poverty line involves judgements about physiological requirements, which vary depending upon individual human characteristics such as age, sex and the level of physical exertion. Other definitions of absolute poverty are based on a consensual level of income which is required to maintain human existence. Examples of such absolute definitions of the poverty line include the 'dollar-a-day' (now US$1.25/day) and 'two-dollars-a-day' criteria which are discussed in this chapter. Concepts of absolute poverty have been used very widely in international comparisons of poverty. Because the real value of the poverty line (i.e. the physiological element of the definitions) remains constant over time, it is possible to make judgements about the extent of poverty reduction based on these absolute measures.

On the other hand, 'relative poverty' is defined with respect to what might be defined as a 'comparator group', and is therefore closely related to concepts of 'relative deprivation'. For example, the comparator might be what is considered as an acceptable standard of living within a particular country at a particular point of time – involving moral or ethical judgements. The criteria for such an acceptable standard of living in any one country are likely to change over time as the general standard of living in that country increases with economic growth – so that the relative poverty line will rise together with the country's standard of living. A typical 'relative' poverty line might be defined as 50 per cent of the median income per capita, which is explicitly linked to the prevailing income distribution.

> For the European Union, the current definition of relative poverty is 60 per cent of the median income per capita: 'In practice, the main measure of poverty used in the EU at present is the Eurostat definition: the percentage of people with an income of 60% or less of the median income in the country in which they live. Although this means that the poverty line, in terms of absolute values, differs between countries, it is indicative of relative deprivation in the country concerned' (European Union, 2009). The relative poverty concept is used in the construction of the UNDP's Human Poverty Index for higher-income countries (HPI-2): 'deprivation in a decent standard of living is measured by the percentage of people living below the income poverty line, set at 50% of the adjusted median household disposable income'. The detailed methods for the calculation of the UNDP's poverty measures are included in Technical Note 1 to the 2007-8 *Human Development Report* (UNDP, 2008: 355–61).
>
> Because the relative poverty line rises as incomes increase, the implication is that relative poverty can be reduced, but only if the income distribution becomes more equal. If the income distribution in a country becomes more unequal, then relative poverty is likely to increase. In this sense, although absolute poverty is likely to be reduced over time as incomes rise, relative poverty is always likely to be with us unless income inequality falls significantly.

ability or 'capability' to achieve valued 'functionings' – valued outcomes or 'ends'. A 'process' freedom refers to one's ability to be an agent and affect processes at work in one's life.

The Capability Approach is the basis for the UNDP's 'Human Development' paradigm including the *Human Development Indices* and other related definitions of development as 'the process of enlarging people's choices' (UNDP, 1990: 1). In the case of poverty assessment and 'basic capabilities', Sen (1992: 44–5) noted that:

> [i]n dealing with extreme poverty . . . [capabilities might include]
> . . . the ability to be well-nourished and well-sheltered, . . .
> escaping avoidable morbidity and premature mortality, and so
> forth.

However, the actual identification of sets of 'capabilities' and 'functionings' remains unresolved after two decades. Indeed, there are now numerous 'capabilities' sets. Nussbaum is also credited with extending Sen's work. Various writers have proposed sets of capabilities (see in particular, Alkire, 2002; Doyal and Gough, 1991; Ekins and Max-Neef, 1992; Nussbaum, 2000).

The Human Development Report (HDR) and the Human Development Indices launch (and the related indices which are discussed below) played a role in what was to become known as the decade in which poverty and social development would rise to prominence in academic and policy arenas.

In the same year as the HDR launch, 1990, the World Bank also issued a new, albeit money-metric or economic measure of poverty, the dollar-a-day poverty indicator (updated in 2009 to US$1.25/day). Throughout the decade there were numerous United Nations conferences many dealing closely or directly with poverty and human development.

The multi-dimensional approach to poverty 'is so common place it is easy to forget it was not always the case' (Kanbur, 2001: 1085). The contribution of development economics has been significant is this. The 2000/1 *World Development Report* (World Bank, 2000) played a role in solidifying developed country support for poverty reduction in the development discourse and promoted a multi-dimensional model of poverty. However, of far greater significance was the UN Millennium Assembly in New York held on 18 September 2000 at which all countries signed the Millennium Declaration from which the Millennium Development Goals – the UN poverty reduction targets for 2015 – are derived (for further details refer to Box 9.2).

Other approaches to conceptualising poverty are livelihoods, rights, exclusion and, most recently, wellbeing (see Box 9.3). All have an economic dimension – economic livelihoods, economic rights, economic exclusion and economic wellbeing. Different frameworks identify different people as 'poor' and imply different responses. In recent years, a relatively new fault-line has emerged – that between universal or 'objective' conceptualisations of poverty (such as those previously discussed above) or 'subjective' or local experiences of wellbeing. This approach is associated in particular with Chambers (1983, 1997, 2006) who argues that the perceptions of poor people (rather than of rich people or members of the development community) should be the point of departure because top-down understandings of poverty may not correspond with how poor people themselves conceptualise changes in their wellbeing. Security, dignity, voice and vulnerability may be more important than consumption. For example, Kingdon and Knight argue that:

Box 9.2

The Millennium Declaration and the Millennium Development Goals

The MDGs are the UN poverty targets for 2015. There are eight MDGs as follows:

MDG 1. Eradicate extreme poverty and hunger
MDG 2. Achieve universal primary education
MDG 3. Promote gender equality and empower women
MDG 4. Reduce child mortality
MDG 5. Improve maternal health
MDG 6. Combat HIV/AIDS, malaria, and other diseases
MDG 7. Ensure environmental sustainability
MDG 8. Develop a global partnership for development

In total eight MDGs include twenty-one quantifiable targets that are measured by sixty indicators.

The MDGs are drawn from the UN Millennium Declaration which was based on six 'fundamental values'. These were freedom (incorporated into MDG 1, 2, 3, 4, 5, 6); equality (MDG 2); solidarity (MDG 8); tolerance (no MDG), respect for nature (MDG 7) and shared responsibility (MDG 8).

The MDGs themselves were in fact a collation of goals agreed at various UN conferences in the preceding period. Support from developed countries and from the Bretton Woods institutions was greatly encouraged by the OECD, which had agreed on many of these targets as part of a high-level meeting (OECD, 1996). Indeed, the MDGs were known in a former guise as the OECD International Development Targets.

Source: United Nations (2007).

an approach which examines the individual's own perception of wellbeing is less imperfect, or more quantifiable, or both, as a guide to forming that value judgment than are the other potential approaches.

(2004: 1)

These psychological elements of development have shifted discussion from objective wellbeing to subjective wellbeing and from physiological conditions (namely the objective physical condition of the individual) to happiness and psychological experience (the subjective psychological experience of the individual). Participatory Poverty Assessments (PPAs) have sought to elicit such perspectives

Box 9.3

Livelihoods, rights, exclusion and wellbeing

The Sustainable Livelihoods approach is associated with the seminal paper
of Chambers and Conway (1992). A 'livelihoods' approach is concerned
with the evaluative space of a 'livelihoods strategy' which is shaped by five
household assets, the context, the institutional rules, norms, etc., and the
policy regime that mediates capital accumulation across those five assets.
The approach can be defined as:

> the process of identifying the resources and strategies of the poor, the
> context within which they operate, the institutions and organisations
> with which they interact and the sustainability of the livelihood
> outcomes which they achieve, providing a way of picking a path
> through this complexity at micro level.
>
> (Shankland, 2000: 6)

A *rights-based approach* (RBA) is not associated with any author(s) in
particular though for discussion see Maxwell (1999). It is associated with
various declarations and conventions of the United Nations. In addition to
the Universal Declaration of Human Rights (UDHR), there is also the
Convention on the Rights of the Child and the Convention on the
Elimination of Discrimination Against Women. An RBA is concerned with
achieved 'rights'. These 'rights' are universal and objectively defined by
various United Nations agreements, in particular the UDHR. These include
the right to food, shelter, education, health care and so on and thus there is
much resonance with the Human Development and Capability Approach.
The RBA provides a set of indicators or 'goals' of development that have
been accepted by all countries in the UDHR and other United Nations
agreements. The RBA shifted the focus to the largest deprivations. It can
be defined as:

> the achievement of human rights as an objective of development . . .
> (invoking) the international apparatus [of] rights accountability in
> support of development action.
>
> (Maxwell, 1999: 1)

The *social exclusion approach* pushed the concern with the most deprived
further by focusing on specific groups. It is concerned with the structural
processes and agency factors that lead to exclusion through group dynamics
of certain groups (such as ethnic minorities, lower-caste members, the
disabled). This approach is concerned with the evaluative space of processes
leading to exclusion and seeks to focus on relative deprivation. It is an
approach that has been popular in the 'North' but is increasingly used in
southern contexts (see for further detail Maxwell, 1998).

A *well-being approach* is 'what a person has, what a person can do with what they have, and how they think about what they have and can do' (McGregor, 2007: 317). It can be argued that the value added or comparative advantage of a wellbeing lens (over a 'traditional' poverty lens) is that it addresses what people feel (their emotions and experiences) as well as what they can do and be. It is about what people can do, be and feel rather than what they cannot. It expands the focus from the body/physiology to include mind/psychology. It is endogenous or based on self-determination/participation rather than exogenous labelled or decided and imposed externally.

on wellbeing from poor households (albeit with the contradiction of having to use some definition of poverty to identify the poor sample beforehand).

The largest study to date has been the World Bank's *Voices of the Poor* (Narayan et al., 1999) which included 60,000 people in more than sixty countries. The study concluded that the poor define poverty as multidimensional including material wellbeing (food security and employment were highlighted), as well as participation and voice in decision making that affects one's life and vulnerability (see Box 9.4).

This leads to a differentiation between the transient poor and the chronic poor. The transient poor move in and out of poverty over the course of a year. The chronic poor are always poor and include those who face (Hulme et al., 2001):

- life cycle deprivations – old, young, widows;
- discrimination owing to minority religion, caste, refugee, indigenes, migrants, etc.;
- disadvantage within households – girls, daughters in law, children where there are many other children;
- chronic health and disability;
- residents of remote areas, urban ghettos, areas of conflict.

As Sen (1999: 4) notes, 'capabilities that adults enjoy are deeply conditional on their experiences as children'. For example, the priority period for nutrition is while the child is in the womb and up to 18 months of age. Malnutrition losses in this period are irreversible – they represent losses the child will carry throughout life. Of the female babies that survive, the ones that remain malnourished in

---

Box 9.4

Vulnerability to poverty

For various reasons, such as seasonality for example, some people and households move in and out of poverty over time – that is, poverty may be a transient phenomenon. This is sometimes referred to as 'churning'. Whilst for others, in either depth, severity or longevity the experience of poverty (by whatever dimension) may be a long-term phenomenon. Vulnerability and poverty are overlapping but different concepts. Poverty itself is about deprivation in various dimensions. Vulnerability is about the risk or probability of an individual, household or community moving in or out of poverty (i.e. transient or chronic poverty by whatever definition taken of poverty) in response to shocks and fluctuations. What kinds of shocks and fluctuations? Such shocks may be environmental in nature (changes related to climate change, for example), economic or market based (changes in access to finance, for example), political risks (changes in workers' rights or conflict in society), social risk (changes in social protection). In short, vulnerability and risk are not just 'new' dimensions of poverty but actual causes of poverty and deprivation in themselves. Concern is with who is at risk, what are the sources of risk and how risk might be insured against.

Source: Authors.

---

adolescence are more likely, in turn, to give birth to malnourished babies. This also raises the issue of the inter-generational transmission of poverty (see Table 9.1). Poverty can be transmitted from parents to their children by assets (see table). According to McKay and Lawson (2003) the transmission of poverty between generations is effected by economic and non-economic factors as shown in Table 9.1 and as follows:

Economic factors:

- Economic trends and shocks (e.g. commodification, shifts in terms of trade, hyperinflation).
- Access to and nature of markets – for example, nature of labour market (employment opportunities for children, young people and women; labour migration as livelihood strategy); access to financial market.
- Presence, quality and accessibility of public, private, and community-based social services and safety nets.

Table 9.1 *The intra-generational transmission of poverty-related capital from 'parent' to 'child'*

What is transmitted?	How is it transmitted?
Financial, material and environmental capital • Cash • Land • Livestock • Housing and other buildings • Other productive/non-productive physical assets (e.g. rickshaw, plough, sewing machine, television) • Common property resources • Debt • Inter vivos gifts and loans	• Insurance, pensions • Inheritance, bequests, dispossession • Dowry, bridewealth • Environmental conservation or degradation • Labour bondage
Human capital • Educational qualifications, knowledge, skills, coping and survival strategies • Good mental and physical health • Disease, impairment • Intelligence?	Socialisation • Investment of time and capital in care • Investment of time and capital in education and training • Investment of time and capital in health, nutrition • Contagion, mother-to-child transmission • Genetic inheritance
Social, cultural and political capital • Traditions, institutions, norms of entitlement, and value systems • Position in community (i.e. family, 'name', kin group, caste, race, nationality, language, physical appearance) • Access to key decision makers, political patrons, civil society organisations and development agencies • 'Culture of poverty'?	Socialisation and education • Kinship • Locality • Genetic inheritance

Source: adapted from McKay and Lawson (2003).

Non-economic factors:

● Norms of entitlement determining access to human capital, particularly education, health care and nutrition.
● Structure of household and family, including headship as well as gender, birth position, marital status and age of 'child' and 'parent'.

- Child fostering practices.
- Education and skill level of 'parent'.
- Intent/attitude of 'parent' and 'child'.
- HIV/AIDS pandemic; other diseases regionally endemic; associated stigma.
- Nature of living space – for example, security/conflict/violence, stigma, remoteness, sanitation.

Economists and poverty reduction policies

Development economics has also made a major contribution to debates on poverty with regard to policy/strategy in a more traditional area of economists' interest – growth – and specifically the relationship between poverty and growth. This contribution is based around the following question: is growth good for the poor? If the answer is yes (either in relative or absolute terms – see below), then there is no need for government policy to depart from a focus on growth strategy. However, if the answer is no, then quite different policies may be required (Bourguignon, 2003; Foster and Székely, 2002).

Although this issue is still contested, there is *some* consensus. Even Dollar and Kraay, the authors of the widely cited study 'Growth is Good for the Poor' (Dollar and Kraay, 2002), distanced themselves from claims, for example by Wade (2001a: 1440 fn.3), that the study represented a manifesto for 'growth-is-all' as well as more popular portrayals in the media. Other contributions on the growth–poverty association have been by Eastwood and Lipton (2001) and by Fields (2001). Indeed, the PRSP process reflects the fact that few people are currently arguing for strategies that are based entirely on growth maximisation (OECD, 2001: 18; UNMP, 2005: 4–9; World Bank, 2000: 45–59).

As Kanbur (2001: 1) noted 'there are few people who argue economic collapse and stagnation is good for the poor. . . . It's the policies that are under debate not growth per se.'

In general, debates within development economics have been instrumental in this shift away from development policies based entirely on economic strategies – that is, growth maximisation – towards recognition of the need to focus on poverty (and inequality) in their income and, more recently, their non-income dimensions as well as 'pro-poor growth'. Klasen (2005) has discussed the non-income dimensions of poverty and their relationship with growth fully. The genesis of these debates can be seen as early as the 1970s, when development economists' concerns that growth was by-passing the poor led to the emergence of calls for 'growth with redistribution'.

Chenery et al. (1974: xiii) argued that a decade of growth had bypassed with 'little or no benefit' a third of the population in developing countries and Adelman and Morris (1973: 189–93) wrote of hundreds of millions 'hurt' by economic development.

In the 1990s, an old idea in new clothing became one of the most pervasive and widespread phrases in development policy: 'Pro-poor growth' (Kraay, 2004). This emerged as a synthesis of 'growth with redistribution', 'broad-based growth' and 'growth with equity'. 'Pro-poor growth' has been defined in numerous ways, but two categories can be outlined in terms of outcomes: those based on whether the poor have benefited in an absolute way – so that the headcount poverty measure falls or the incomes of the poor rise – and those based on the poor benefiting in a relative sense that implicitly entails reductions in inequality (see in particular, Ravallion, 2004; Ravallion and Chen, 2003; White and Anderson, 2001).

The current consensus is that growth is important but inequality matters too (Maxwell, 2005: 2, 4; OECD, 2001: 18; World Bank, 2000: 45–59). The concept of the 'poverty elasticity of growth' has been the subject of numerous studies and it measures the relationship between poverty and growth. A higher value of the poverty elasticity of growth means greater poverty reduction for any given rate of economic growth. A lower value for poverty elasticity means that there will be less poverty reduction for any given rate of economic growth. The poverty elasticity of growth varies widely across countries and across time within countries (White and Anderson, 2001; Eastwood and Lipton, 2001; Ravallion, 2004).

One viewpoint is that growth *is* enough and that significant poverty reduction occurs as a result of economic growth, in which case there is no need for 'pro-poor growth' because the incomes of the poor rise one-for-one in line with average income. This amounts to what is usually described as the 'trickle-down effect'. This has been the subject of several cross-country econometric studies (Dollar and Kraay, 2002; Gallup et al., 1999; Roemer and Gugerty, 1997), and an associated finding has been that the poverty headcount ratio declines significantly with growth (Bruno et al., 1998; Ravallion, 1995, 2001; Ravallion and Chen, 1997). However, many of these, notably the Dollar and Kraay study, have faced sustained criticism on methodological grounds (see Amann et al., 2006; Nye et al., 2002; Weisbrot et al., 2001).

Surprisingly perhaps, growth has even been associated with *increases* in poverty in much of sub-Saharan Africa, Russia and much of Eastern Europe (Epaulard, 2003; Mosley, 2004).

Variance in poverty elasticities of growth across countries has brought to light the role of inequality, thus challenging the 'growth-is-all' perspective. Initial inequality has most commonly been identified as deterministic within the heterogeneity of country experience: a higher level of inequality leads to less poverty reduction at a given level of growth (Deininger and Squire, 1998; Ravallion, 2004). Bourguignon (2003) argued that overall, half of poverty reduction was due to growth effects and half to distribution. White and Anderson (2001) found, in a quarter of 143 growth 'episodes', that the distribution effect was stronger than the growth effect.

Why might countries differ so markedly? As Mosley (2004: 7) has noted, there are a large number of potential variables to choose from. The heterogeneity of country experience has also been linked to changes in inequality over time due to geographical differences (urban–rural); the composition of public expenditure; labour markets; social capital endowments and the variance in actual rates of growth (Mosley, 2004; Mosley et al., 2004). The fact remains that growth is often likely to be unequal. What can policy makers do to redistribute the benefits of growth? Box 9.5 outlines three important areas of policy options.

Given that poverty is now thought of as multi-dimensional, development economics needs to revisit much research on income poverty and growth. Growth is clearly not an end in itself, but rather a means to other ends. Growth definitely does supply essential resources for the attainment of these ends, through both private and public channels, if the benefits of growth can be sufficiently widely shared.

To echo Sen (1999), income is *only* an 'instrumental' freedom (i.e. it helps to achieve other 'constitutive' freedoms such as being healthy or being well fed). The key question then is how growth relates to these

Box 9.5

Policies for poverty reducing growth

Redistributive and transformative public expenditures to break the inter-generational transmission of poverty

Policy can redistribute the benefits of growth through pro-poor public expenditure. Growth is a major potential source of government revenue to finance public expenditure, which can be designed to be explicitly pro-poor – for example, through broad-based expenditure on education and health. This provides an important opportunity for the benefits of growth to be more widely shared, and in a manner which is not likely to have major disincentive effects that would crowd out future growth. On the contrary, increased spending on education, nutrition and health, as well as key items such as infrastructure, is likely to be an important basis for future growth. As part of this, investment in young children and their families, via nutrition, health or education programmes – for example, in order to break the widespread intergenerational transmission of poverty – potentially offers very high returns. It remains, though, always a major challenge to make sure that public spending is not captured by the rich.

Increasing the rate of job creation from growth

It is also important that growth is associated with significant job creation to provide opportunities to people to benefit from higher education levels and move out of agriculture. But the record of employment creation with growth has been very weak in many countries. How can policy increase the job creation from growth? Increased levels of private sector investment is one important part of the story, and that is likely to require substantial financial sector development. There is also potential for job creation through more informal channels by reducing formal entry requirements and rules on informal sector trading, as well as investment in small-scale infrastructure.

Broad-based sectoral growth, particularly supporting food crop agriculture

Job creation may not benefit the poorest directly. Therefore, it is highly desirable to have a pattern of growth which is broad based in terms of its coverage of sectors, regions or population, including the agricultural sector if that is the sector in which the poor are disproportionately represented. Investment in market development, research, infrastructure and value-added processing activities may all be important. Fast agricultural growth may also form a basis for transformative growth with the sectoral composition of

continued

growth shifting towards manufacturing and services. Investment in social protection (measures to reduce vulnerability to poverty) can also potentially play a major role by reducing the vulnerability of small farmers and the poor in general.

Source: McKay and Sumner (2008: 4).

ultimate ends. If we consider that ultimate ends are concerned with human development, reduction of vulnerability, participation, psychological well-being etc., then we need to understand how the growth process interacts with these. There are serious gaps in knowledge on these questions. However, this should not detract from the fact that development strategies based entirely on economic considerations, with only limited attention to poverty and inequality, are largely a thing of the past with development economics having played an important role in this change.

9.3 Selected contributions by development economists to the analysis of inequality

Economists and the meaning of inequality

Development economics has made a major contribution to debates on the meaning and measurement of income inequality, based for example on the Gini-coefficient (see Box 9.6a). Other contributions include Atkinson's (1970, 1983) seminal work on the measurement of inequality and the work of Bourguignon (1979) and Shorrocks (1983).

In the context of development studies, development economics has linked the measurement of income inequality directly to poverty (Foster et al., 1984; Sen, 1976) and the poverty 'gap' and poverty 'severity' measures (see Box 9.6b).

Poverty is more severe in a household that is 30 per cent below the poverty line as compared to a household that is only 5 per cent below the poverty line (reflected in the poverty severity index), although the headcount measurement would not differentiate between these two households. The poverty gap shows the intensity of poverty, through

Box 9.6a

The Lorenz curve, and the Gini-coefficient

The Lorenz curve is a frequency distribution which describes the income distribution. The percentage of income is shown on the vertical axis and the percentage of population (or of households) is shown on the horizontal axis. In the diagram below, the diagonal line connecting the top right-hand and bottom left-hand corners is a line of perfect equality – 10 per cent of the population receives 10 per cent of the income, 90 per cent of the population receives 90 per cent of the income – and so on. The Lorenz curve shows the actual income distribution, and the closer it is to the right-hand vertical axis and the bottom horizontal axis, the greater the degree of income inequality. So, for example, the hypothetical Lorenz curve in the diagram shows that the poorest 20 per cent of the population receive approximately 5 per cent of the income, and the richest 5 per cent of the population receive approximately 20 per cent of the income. This is directly relevant to the measurement of the economic or income dimension of poverty.

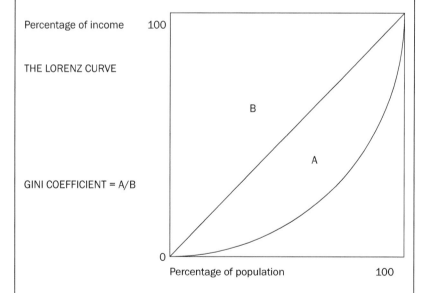

The Gini-coefficient gives a numerical value to the degree of inequality of the income distribution and is given by the ratio of the area between the diagonal and the Lorenz curve (A) and the area between the diagonal and the vertical and horizontal axes (B). The closer the Lorenz curve is to the axes, the higher the value of the Gini-coefficient, with the limiting values

continued

being 0.0 for complete equality and 1.0 being complete inequality. For example, the Gini-coefficient in 1990 for Brazil was 0.607 whilst for Costa Rica it was 0.457 with comparable figures in 2000 for South Africa and the United States being respectively 0.578 and 0.408 (World Bank, 2007a). There are a number of theoretical and practical difficulties in securing reliable estimates for the Lorenz curve and for the Gini-coefficient, which partially accounts for the sparse nature of the available data.

measurement of the difference between the average income of all poor households and the poverty line. These measures emerged out of a development economics debate on the identification of 'axioms' – what conditions should a poverty measure demonstrate? For a fuller discussion Ravallion and Chen (2003), Atkinson (1987) and Zheng (1993) are very useful sources.

The intra-household allocation of income, consumption and resources is another important dimension of inequality. Much poverty analysis and data is at the household level, and assumes income and other resources are equally shared. In reality, this assumption often does not hold. For example, malnutrition, illiteracy and other dimensions of poverty may be significantly worse amongst women. Children and girls in particular may suffer hidden deprivations and greater depth of deprivation. This may be hidden in the sense that 'traditional' proxy monetary measures of poverty and sources of data, such as income and consumption, are deeply problematic for children because:

- data are not collected from women or children themselves but male heads of household and in the case of children, carers;
- women and children's employment may be in the informal economy;
- non-market channels may be more important in shaping gender dimensions of poverty and childhood poverty;
- women and children's access to and control of income may be extremely marginal and resources and power are distributed unequally within the household.

Box 9.6b

The poverty gap

The poverty gap approach to the measurement of poverty relates to the question of what percentage of national income would have to be re-distributed from those above the poverty line to those below in order to bring them up to the poverty line. The proportion of the population below the poverty line and the size of the poverty gap will depend upon the level of the poverty line (e.g. $1 a day or $2 a day) and the shape and position of the curve showing the income distribution. The poverty gap measure therefore gives an indication of the 'depth' of poverty.

One of the most significant criticisms of the Dollar–Kraay analysis of the relationship of economic growth to levels of poverty is that it defined those in poverty as being the lowest quintile of the income distribution (the lowest 20 per cent) without taking into account the distribution of income within that quintile. The poverty gap measure avoids this pitfall.

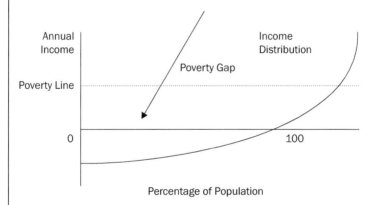

One easily accessible source for a description of the poverty gap is Todaro and Smith (2006: 204–5).

Economists and inequality policy

The issue of inequality and growth is associated with one of the best known and most controversial theories in development economics, the Kuznets curve. Simon Kuznets (1955; 1963) introduced this in his presidential address to the 1954 American Economic Association.

In this address, and in later articles, he set up a 'hypothetical numerical exercise' which he noted was based 5 per cent on empirical information and 95 per cent on speculation (cited in Kanbur and Squire, 2001: 192). Kuznets postulated an inverted U-shaped relationship between income and inequality, and predicted an increased inequality in the early stages of development followed by reduced inequality in subsequent periods. This approach is consistent with the Lewis dual economy model (see Chapter 2). Jalilian and Kirkpatrick have recently explored this approach (2005: 639–40). Lewis, however, did not originally assume that a rise in inequality would be inevitable.

Kuznets argued that agricultural economies (for example, those of developing countries) are relatively equal societies but with low average income. As the economy develops, the population gradually shifts to non-agricultural sectors, where average incomes are higher, leading to increased overall inequality. Thus in the early stages of development, inequality increases because of the low but rising proportion of national income in the modern industrial sector and the rising proportion of profits in national income. In time, as more of the population moves out of the traditional, rural, agricultural sector into the modern, urban, industrial sector and real wages in industry begin to rise, income inequality decreases. Greater elaboration, and a review of theoretical literature on the Kuznets curve, can be found in Deutsch and Silber (2004) in particular.

The dominant view is that inequality is not an outcome of growth but that it plays a role in determining the evolving patterns of growth and poverty reduction (Bourguignon, 2003: 12). The fact that there are many country deviations from 'average experience' means that the relationship between economic growth and income inequality/distribution is not one for easy generalisations. Economic growth can impact on inequality through a variety of channels including modification to the distribution of resources across sectors, and changes in relative prices, factor rewards or factor endowments. Deininger and Squire note (1998) that failure to find the Kuznets curve relationship overall does not mean it does not exist for individual countries. In four countries in their forty-nine-country sample, the Kuznets hypothesis was supported. However, most studies have focused not on the relationship which Kuznets hypothesised from growth-to-inequality but rather on the implied trade-offs which might exist in the inequality-to-growth relationship (see Table 9.2). Much

Table 9.2 *Summary of empirical work on inequality-to-growth relationship*

	Positive association – inequality is good for growth	*No association*	*Negative association – inequality is bad for growth*
The impact of initial inequality on future growth	*Early studies*: Adelman and Morris (1973), Ahluwalia (1976), Ahluwalia et al. (1979), Paukert (1973)	Bruno et al. (1998), Deininger and Squire (1996), Knowles (2005), Li et al. (1998), Lopez (2005)	Alesina and Rodrik (1994), Birdsall et al. (1995), Clarke (1995), Easterly (2002), Perotti (1996)
	Later studies: Deutsch and Silber (2004), Forbes (2000), Li et al. (1998).		*Asset (land) inequality*: Birdsall and Londõno (1997), Deininger and Squire (1998)

Note: unless stated inequality refers to income/expenditure inequality.

of the debate has focused on the latter perhaps because no systematic association from growth-to-inequality has been reported in the recent empirical work (see, for example, Adams, 2003; Deininger and Squire, 1998; Dollar and Kraay, 2002; Easterly, 1999a; Ravallion and Chen, 1997). Indeed, many studies have argued that, year-to-year, intra-country inequality does not change a great deal (Deininger and Squire, 1998; Gallup et al., 1999; Li et al., 1998; Ravallion, 2001; Roemer and Gugerty, 1997; Timmer, 1997).

A number of studies in the 1970s initially supported the contention that initial inequality had a positive impact on subsequent growth. However, in the 1990s a series of new studies led by Anand and Kanbur (1993a, 1993b) questioned this approach. Some argued that there was no empirical relationship between initial inequality and growth or that greater inequality actually reduces subsequent economic growth. Some other studies have found that redistribution was good for growth (Easterly and Rebelo, 1993; Perotti, 1996). Most recently, the empirical literature has become somewhat contradictory. Some have reaffirmed the positive impact of inequality on growth whilst others have found the exact opposite using asset inequality as well as income inequality (see Table 9.3). Yet others have posited that

Table 9.3 *Most commonly used 'economic' and 'non-economic' poverty measures*

Economic Poverty Indicators

Income per capita
- GDP per capita (PPP – purchasing power parity)
- Real wages
- Unemployment rate*

Income poverty lines
- Percentage of the population living under a dollar–a–day per capita*
- Percentage of the population living under the national poverty line
- Percentage of the population vulnerable to poverty through variance of income or assets
- Income inequality
- Poverty gap and severity indices at a dollar–a–day per capita*
- Expenditure of bottom quintile as percentage of total expenditure*
- Gini-coefficient

Non–economic Poverty Indicators

Education
- Education enrolment rates*
- Survival to the final primary or secondary school grade/completion of primary or secondary school*
- Literacy rates*

Health and nutrition
- Malnutrition rates*/food or calorie consumption per capita/Body mass index
- Mortality and morbidity rates*/life expectancy/not expected to survive to forty years/infection rates*
- Health service usage – skilled personnel at birth*/contraceptive prevalence rate*/immunisation rates*

Environment
- Access to 'improved' water sources*
- Access to 'adequate' sanitation*
- Household infrastructure – permanent material used for walls of home and electricity supply

Note: * denotes an MDG indicator.

Source: Author.

the inequality–growth relationship depends on the level of economic development (Barro, 2000; List and Gallet, 1999) or differs between democratic and non-democratic countries (Deininger and Squire, 1998; Perotti, 1996) or that any reduction in inequality reduces future growth (Banerjee and Duflo, 2003; Forbes, 2000). However, Klasen (2005) argues that these studies have too few observations to be reliable.

A notable issue is the nature of the raw dataset. The original Kuznets curve was based on historical data from the first half of the nineteenth century for three developed countries (USA, UK and Germany), although some reference was also made to India, Ceylon (now Sri Lanka), Prussia and Puerto Rico. The 1970s studies were primarily based on cross-sectional rather than time-series data (problematic for estimating inter-temporal relationships) and many of the 1990s studies have been based on the new, larger, 'high-quality' dataset constructed by Deininger and Squire.[1] Additionally, as Knowles (2005: 139) notes, studies based on short-run periods find that the Kuznets inequality-to-growth trade off is supported (for example, Forbes, 2000; Li et al., 1998) but long-run studies do not (for example, Alesina and Rodrik, 1994; Birdsall et al., 1995; Easterly, 2002).

Some revisiting of conceptual ideas has also been evident relating to how initial inequality reduces subsequent growth. Greater detail can be found in Aghion et al. (1999), but the arguments will be summarised here. The first 'new' theory has been labelled the 'redistributive political-economy model' (see Rehme, 2001). This is based on the idea that unequal societies create redistributive pressures leading to distortionary fiscal policies that reduce future growth. Empirical evidence on these issues is mixed and a range of views can be found in Clarke (1995), Deininger and Squire (1998) and Perotti (1996). Second, a 'socio-political instability' theory (see Alesina and Perotti, 1996) has suggested that income inequality creates socio-political instability which leads to reduced investment and hence lower economic growth. Third, an 'imperfect credit markets' theory (see Ravallion, 1998) is related to the notion that in unequal societies there is a high density of credit-constrained people leading to lower investment (especially in human capital) and hence lower economic growth.

Given that inequality is now also thought of as being multi-dimensional, development economics needs to revisit much of the research on inequality and growth. However, this should not detract from the enormous contribution made by development economics to inequality analysis.

9.4 Indicators of poverty and inequality

The purpose of this section is to illustrate that the process of constructing satisfactory poverty and inequality measures, in particular ones that do not send perverse signals to policy makers, is highly problematic. At a first stage, one can consider: what characteristics does a 'good' poverty indicator exhibit? There is little disagreement on the characteristics of a 'good' indicator. Most commonly noted are the following criteria – the measure should:

- be 'user friendly' – that is, be relevant/useful to policy/decision makers;
- be relatively cheap, easy to collect and not easily manipulated;
- have an underlying universal conceptualisation of poverty (and thus legitimacy) – that is, the same to all people, at least locally, if not globally (the World Health Organization estimates human beings need food on average as a minimum of 2100 calories for example);
- be a simple, direct, measurable, unambiguous proxy rather than an indirect/ambiguous proxy.

Sources: DFID (2005), Maxwell (1999),
World Bank (2002).

This is fine in theory, but what commonly used poverty indicators could jump through these series of 'hoops'? A second stage is the question of different poverty meanings and dimensions. Who the poor are (Sen's identification problem), how many 'poor' people there are (Sen's aggregation problem), and the severity of poverty are critically dependent on the choice and ranking (and thus weighting of dimensions) in indicator(s). Approximately one billion people live under a dollar-a-day but over two billion lack access to 'adequate' sanitation (UN, 2007). Which is more important? Should they be equally weighted or ranked? There is no consensus on which dimensions deserve greater emphasis/priority. If the number of poor or the percentage of poor people falls by one dimension/measure – for example, literacy – but increases by another dimension/measure – for example, life expectancy – has poverty overall risen or fallen?

Further, not only can individuals be poor in some dimensions but not poor in others, but poverty is also dynamic and people move in and out of poverty over time – sometimes over relatively short periods. A third stage is the question of what policy signals do different indicators send?

Indicators of economic and non-economic poverty and inequality

'Economic' poverty indicators can be split into groupings: those loosely based on a) income per capita, b) those utilising an income poverty line, and c) those assessing income inequality (see Table 9.3). 'Non-economic' poverty measures can also be grouped into three clusters (see again, Table 9.3): those based around a) education; b) health and nutrition, and c) the environment. Here the merits of 'economic' versus 'non-economic' measures are discussed rather than individual indicators.

It is not to the exclusion of 'non-economic' indicators but 'economic' measures do have some higher status with policy makers as illustrated in the MDGs, the HDI and the PRSP process. Why might 'economic' measures have this 'preferred' status? There are technical reasons. As noted before, policy makers seek 'user friendly' – that is, relevant/useful to policy/decision makers – indicators that are relatively cheap and easy to collect. 'Economic' measures of wellbeing are popular with policy makers because they are useful when quick, rough-and-ready, short-run, aggregate inferences are required to make an assessment. They are more responsive, changing much faster than 'non-economic' social data (that suffer a time lag). They are more likely to be recently available than 'non-economic' measures and are also cheaper and less complex to collect than 'non-economic' poverty data.

In contrast, 'non-economic' measures of poverty are more useful when a medium- or longer-run assessment is required, because they more directly address the 'ends' or outcomes of policy (being educated and healthy) rather than the inputs or 'means' (greater income). Although they are slower and more expensive to collect (often requiring their own tailored surveys and/or combined methods), they have the additional benefit of being amenable to disaggregation, making them instructive for distributional impacts of policy changes (World Bank, 2001a, 2001b). Given the flaws in both 'economic' and 'non-economic' poverty indicators, a pertinent question might be: do composite measures make up for deficiencies or exacerbate them?

Composite indicators

There is a plethora of composite poverty measures (see extensive review of OECD, 2007). However, the most widely used composites are the UNDP Human Development Indices – the Human Development Index (HDI), the Gender Development Index (GDI) and the Human Poverty Index (HPI). Table 9.4 outlines the components of each. The HDI, GDI and HPI each take account of wellbeing related to longevity of life and health, knowledge and education, and standard of living. There is also a Gender Empowerment Measure (GEM). The GEM is a measure of gender equality in politics, business and wages. The HPI is available in two composite indicators. The HPI-1 is listed in Table 9.4 and there is also an HPI-2 measure that focuses on a selection of high-income OECD countries. These HPI measures are concerned with the measurement of relative poverty (how individual standards of living relate to the country average standard of living) rather than with absolute poverty. The concepts of relative and absolute poverty are explained further in Box 9.1.

Table 9.4 *UNDP composite indicators and their components*

Component indicators

	Longevity	*Knowledge*	*Decent standard of living*
HDI	• Life expectancy at birth	• Adult literacy rate • Combined enrolment rate	• Adjusted income capita (US$ PPP)
GDI	• Female and male life expectancy at birth	• Female and male adult literacy rate • Female and male combined enrolment ratio	• Female and male earned income share
HPI-1	• Percentage of people not expected to live to forty	• Adult illiteracy rate	• Percentage of the population without access to safe water • Percentage of the population without access to health services • Percentage of undernourished children under five

Source: UNDP (various years).

What are the strengths/weaknesses of composites? Generally, they do not tell us how many people are poor or who is poor and there are a number of concerns relating to these UNDP measures – principally, the HDI and GDI show little more than income per capita and that the index components themselves correlate very closely questioning the rationale for the measure. However, it is the fault of the component parts that more seriously undermine their validity – often data do not exist for a particular year, resulting in the nearest available year being used or estimated by UN country staff. For example, given the large gaps in recent health and education data, the 2002 HDI for many countries may be made up of current GDP per capita but with education and health data from the mid-1980s or 1990s when the last survey was conducted or more recent guesses by UN country staff based on GDP per capita extrapolation (for more detailed HDI critique, see McGillivray, 1991). In sum, composite measures cannot make up for the deficiencies in components and most simply emphasise 'economic' wellbeing.

9.5 Summary

- In their 2001 review of the evolution of thinking on poverty and inequality, Kanbur and Squire concluded by asking what Rowntree would have to say if he were alive today? They suggested he might be surprised, one hundred years on, that income was still a measurement for poverty and inequality, but they would be likely to agree that health and education are important factors in wellbeing.
- What might some of the founding fathers of quantitative economics or classical political economy have to say if they were alive today? All would have been likely to emphasise the essential link between poverty and economic welfare.
- However, Smith and Marx might have added that any over-emphasis on the 'economic' dimensions would be to deny the broader aspects of the human condition. Marx might have placed more emphasis than Smith on the corrupting influence of an over-emphasis on the importance of money/income on moral sentiments.
- One might speculate that Quesnay, Ricardo and Malthus would have taken a close interest in the importance of malnutrition, given their shared interest in agricultural output, and Petty might well have focused on the provision of public goods such as health and education given his work on public finance and fiscal policy.

- However, it might well have been John Stuart Mill who would have had the most to say about the current poverty and inequality debates given his focus on the importance of economic, political and social freedoms.
- Contemporary development economics has made a major contribution to debates about inequality with regard to policy/strategy in a more traditional area of economists' interest – growth – and specifically the relationship between income inequality and growth. This discussion has more recently been extended to cover the relationship between asset inequality (particularly land) and growth.
- If income or asset inequality are conducive to relatively high rates of economic growth, then redistributive policies would not be justified on growth grounds. However, if they are not good for growth, then quite different policies are required.
- What has been implied in the past by the inequality–growth linkage is that there is a trade off: inequality is a short-term price worth paying for long-term economic development, and the economic growth which ensues would eventually lead mechanistically to poverty reduction. Considerable doubt has been cast on this proposition. Inequality is not necessarily conducive to growth, and so there may not be a 'trade off', and growth is necessary but not a sufficient condition for poverty reduction.
- Three clusters of poverty/wellbeing indicators can be identified and categorised: a) those that measure poverty as primarily 'economic' wellbeing; b) those that measure poverty as primarily 'non-economic' wellbeing; and c) composites. The commonly used indicators have been drawn from the major annual poverty-related publications – UNDP's Human Development Report, and the World Bank's World Development Report. Many of these measures are utilised in the Millennium Development Goals and thus are of particular interest.

Questions for discussion

1 What is poverty?

2 What is inequality?

3 How are poverty, inequality and growth related?

4 What are the main distinctions between economic and non-economic concepts of poverty?

5 What are the strengths and weaknesses of the main indicators of economic and non-economic poverty and inequality?

Suggested further reading

Besley, T. and Cord, L. (eds) 2006. *Delivering on the Promise of Pro-poor Growth: Insights and Lessons from Country Experiences*. Houndmills, Basingstoke: Palgrave Macmillan.

Grimm, M., Klasen, S. and McKay, A. 2007. *Determinants of Pro-Poor Growth: Analytical Issues and Findings from Country Cases*. Houndmills, Basingstoke: Palgrave Macmillan.

McGillivray, M. 2006. *Inequality, Poverty and Wellbeing*. Houndmills, Basingstoke: Palgrave Macmillan.

Sen, A. 1999. *Development as Freedom*. Oxford: Oxford University Press.

Shorrocks, A. and van der Hoeven, R. 2004. *Growth, Inequality and Poverty: Prospects for Pro-Poor Economic Development*. London and New York: Oxford University Press.

UNDP Annual from 1990. *Human Development Report*. New York: Oxford University Press for the UNDP.

UNDP International Poverty Centre 2006. *What is Poverty? Concepts and Measures*. Poverty in Focus, December. Brasilia: UNDP IPC.

UNDP International Poverty Centre 2007. *Analysing and Achieving Pro-Poor Growth*. Poverty in Focus, March. Brasilia: UNDP IPC.

Economic concepts used in this chapter

Absolute poverty
Basic needs
Capabilities
Composite indicators of poverty
Economic livelihoods
Economic rights
Economic exclusion
Economic wellbeing
GDP per capita
Economic growth
Frequency distribution
Gini-coefficient
Headcount Index
Household disposable income
Human development
Income poverty

Inequality
Kuznets curve
Levels of living
Lorenz curve
Median income per capita
Multi-dimensional poverty
Poverty
Poverty elasticity of growth
Poverty gap
Poverty line
Poverty severity
Pro-poor growth
Relative poverty
Trickle-down effect
Unemployment

Notes

1 The Deininger and Squire (1996) 'higher quality dataset' has since been extended by the United Nations University, World Institute for Development Economics Research (WIDER) in collaboration with the United Nations Development Programme (UNDP) into the World Income Inequality Database (WIID). Further details and a downloadable database can be found at www.wider.unu.edu/research/Database/en_GB/database/

10 Conclusion

10.1 Introduction

The two key themes that have been central to this book are: first, that for a broad understanding of the process of international development (i.e. the subject matter of development studies), it is essential to have some familiarity with the basic economic dimensions of this process; and, second, that there are fundamental disputes between economists with very different theoretical and ideological outlooks, which can lead to confusing messages to non-economists.

The economics of development, or development economics, is a combination of both basic economic tools of analysis (for example, supply and demand analysis) and the specific and often original approaches and tools of analysis that have been developed and popularised over the past fifty or so years by economists specialising in the attempt to gain a greater understanding of the process of economic development. But the 'economics of development' is not a single, unified body of work, and 'development economists' do not have an agreed and universal approach to the fundamental issues of development and considerable differences exist within the profession.

These may seem to be obvious points, but they are very important. Chang (2003a: 3) argues that, up to the 1970s, being a development economist did not necessarily mean being a neo-classical economist, and that even for neo-classical economists it was not necessary to be a supporter of free market capitalism (neo-liberalism). The early development economists (of the 1950s, 1960s and 1970s) were theoretically innovative and remarkably pragmatic (with some notable exceptions), especially with respect to policy prescription. In addition, most would have accepted the need for development economists to have a good understanding of history, politics, society and its institutions and would not have attempted to draw artificial boundaries around their discipline (and to make it more technically difficult through the greater use of mathematics in order to exclude non-specialists).

This situation changed in the 1970s with the advent of the neo-classical counterrevolution (Toye, 1987, 1993) and the rise of market fundamentalism, leading to financial sector and trade liberalisation, deregulation and privatisation, within the context of World Bank structural adjustment and IMF stabilisation programmes. The so-called 'Washington Consensus' has ruled in one form or another for approximately thirty-five years (see Chapter 8), but it is likely that even if the current global financial and economic crisis (late 2008 and 2009) does not signal the end of the 'Consensus', it will inevitably lead to some fundamental policy changes as developing countries struggle to alleviate the impact of the crisis on their fragile economies and as developed market economies recover from the most serious economic and financial crisis since the early 1930s.

The crisis of 2008–9 also calls further into question the extent to which the Millennium Development Goals (Sumner and Tiwari, 2009b) will be achieved by the weaker, more fragile economies. Even before the crisis erupted, it seemed doubtful that the majority of sub-Saharan economies would succeed in reaching them and, as we noted in Chapter 9, a rise in global poverty is now a more likely prospect. But even if the MDGs were to be achieved, that would not in itself represent the achievement of economic development, however defined. Significant poverty remains and structural change and Schumpeterian 'creative destruction' (Nixson, 2006) will lead to continuous upheaval which poor countries are least able to cope with. The future for developing countries is indeed uncertain, given the way the current crisis is unfolding, and it would be foolhardy to make firm predictions.

The other great uncertainty is the impact of the unfolding global environmental crisis on the poorer, more ecologically fragile economies of the developing countries. Development economists have nearly always put economic growth at the centre of their theorising and modelling of the development process, seeing growth as necessary but not sufficient for development. As energy and resource (for example, water shortages) constraints begin to bite, and food insecurity rises, development economists will have to conceptualise a very different development process from the 'traditional' process of growth and change with ever higher levels of per capita (and total) consumption and involving lifestyles that approximate to some 'Western' model of development.

10.2 The 2008–9 financial crisis

As a response to the global financial crisis, the United Nations secretary general appointed a commission of experts, chaired by Joseph Stiglitz, to report on reforms of the International Monetary and Financial System (United Nations, 2009b). The draft report highlights a number of 'basic principles' which should guide the responses of the international community to the crisis. These include:

- restoring balance between the market and government;
- greater transparency and accountability;
- short-run actions consistent with long-run visions (policy actions should not exacerbate the current crisis through their impact on other countries or result in structural changes which increase future instability or reduce future growth);
- assessing distributional impacts (policy makers must be responsive to potential distributional consequences of economic policy actions);
- avoiding an increase in global imbalances and asymmetries (measures introduced to deal with the crisis should reduce, not increase, global inequalities and asymmetries);
- distribution and incidence of risk (under different economic policies, different economic groups bear the brunt of the risks and uncertainties associated with all economic policies).

For the developing countries, the economic crisis will impose huge burdens. The draft report argues that:

- the citizens of poor countries have fewer resources with which to cope with the crisis;

- they already suffer from a lack of automatic stabilisers because of the poorly developed nature of their fiscal and social protection systems;
- 'markets' impose constraints on their ability to pursue countercyclical fiscal and monetary policies;
- countries that have liberalised their financial markets, opened the capital accounts of their balance of payments and relied on private finance from international capital markets, are more vulnerable to current and future economic threats.

These issues take us a long way beyond the content of this modest book, but they illustrate the important role that sensible economic analysis can play in helping us understand developments in the global economy and their impact on the developing countries. We referred to the 2008–9 global financial crisis in Chapter 6 in the context of globalisation. It is not possible at the present time (mid-2009) to quantify the full impact of the crisis on developing countries, but we are able to identify the channels through which the crisis will impact on the development process.

Most obviously, the crisis will impact on the balance of payments of developing countries and through the external accounts will then impact on the domestic economy. In what ways might this happen?

- Falling incomes and declining economic growth in the developed market economies will lead to a fall in the growth of exports of developing countries and a possible fall in export prices.
- It is likely that the flows of capital to developing countries will fall, from both private sources (especially portfolio lending by banks and direct foreign investment by transnational corporations) and official sources (aid). Although donor countries have promised to maintain aid flows, the temptation to reduce them will be great as budgets in donor countries come under greater pressure and the need to reduce budget deficits grows.
- There is likely to be a fall in the flow of remittances to countries that 'export' labour for work overseas. This effect is already happening in a number of countries, for example in Moldova, a country where nearly one-third of the labour force has been working overseas (IOM, 2009; Migration Policy Institute, 2007).
- Reductions in foreign exchange availability will in turn lead to reductions in the imports of developing countries, which in turn will impact adversely on investment and economic growth, thus

adding to the vicious circle – falls in demand for exports, labour, other services, etc. leading to falls in demand for imports, lower growth, less investment (and perhaps less expenditure on health and education), falls in demand for imports in turn feeding into slower global economic growth. The majority of poor countries may be too small economically to have a significant effect on global economic growth, but that is certainly not true for China (and perhaps India), hence the perceived importance of China maintaining a significant rate of economic growth and, indeed, increasing its rate of domestic investment and consumption (that is, reduce its rate of saving) in order to sustain global growth as much as possible.

The only glimmer of hope for developing countries is that they will reassess development policy and move away from the market fundamentalism of the 'Washington Consensus'. Markets, especially financial markets, cannot be totally deregulated without re-regulation to attempt to minimise market failures, improper behaviour and excessive risk taking. Trade policy needs to be radically re-evaluated as Chang (2002) has so effectively argued, and developing countries must once more formulate policies that will focus on the agricultural and industrial sectors, rather than an amorphous and unfocused 'private sector'.

10.3 Globalisation

The likely impact of the financial crisis and consequent economic policy change in developing countries is intimately linked to globalisation. We noted in Chapter 6 that in key respects, globalisation reduces the autonomy of the nation state in key policy-making areas. Most economists would argue, however, that a more active state is needed to deal with the impact of the global crisis on national economies and that there is a clear conflict between the active, interventionist state and the neo-liberal model of globalisation that has dominated mainstream economics and politics for the past two decades.

It is too early to speculate how this conflict will resolve itself in the near future. The emphasis being placed on the role of the IMF, with increased resources devoted to the hardest-hit countries is a cause for concern to many, given the orthodox, deflationary policy stance that the IMF adopts to economic stabilisation. The need for 'sensible' macroeconomic policy (the term used by the eminent North American 'structuralist' economist, Lance Taylor (1988)) is more urgent than ever, and demonstrates both the need for an understanding of economics and a recognition of the variety of different schools of thought.

10.4 The MDGs and poverty alleviation

The MDGs have played a major role in focusing development policy since their inception in the late 1990s (Sumner and Tiwari, 2009a, 2009b). We have already noted that some regions/countries were not on target to achieve those goals by 2015 – for example, most sub-Saharan economies – and the 2008–9 global economic crisis has made their achievement even less likely.

Slower economic growth, other things being equal, will make poverty reduction more difficult (MDG 1). In particular, slower growth in India and China will slow the rate of global poverty reduction. The MDGs dealing with education, health and gender issues (MDGs 2, 3, 4 and 5) are all highly dependent on public expenditure and often on aid flows (Sumner and Tiwari, 2009b), which are both likely to come under increased pressure as the global economic crisis unfolds. As Sumner and Tiwari conclude: 'The evolving context for poverty reduction is likely to be a world which is more uncertain, more complex and more global and within which understandings of poverty itself are evolving' (2009b: 835).

10.5 Climate change, the environment and economic development

It is fair to say that development economists have not yet taken fully into account the likely impact of climate change on the process of economic development. In part, of course, this is because of the complexity and uncertainty of the issues involved. But as a scientific consensus emerges around the work, for example, of the Intergovernmental Panel on Climate Change (2007), we can no longer afford to ignore the likely consequences of climate change, in the form of global warming, on economic and social development.

A recently published report from the Global Humanitarian Forum (2009) argues that climate change is having a disproportionate effect on the world's poorest and most vulnerable populations, many of whom live in harsh environments such as coastal flood areas, desert borderlands, tropical cyclone zones and urban shanty towns. The most critical areas of the global impact of climate change are identified as food, health, poverty, water, human displacement and security. The failure to deal adequately with these problems will lead to both the failure to achieve the MDGs and the wider failure to achieve anything approaching a process of sustainable development.

The findings of the Forum report are stark. It is estimated that every year climate change leaves over 300,000 people dead, 325 million people seriously affected and incurs economic losses of US$ 125 billion. Four billion people are vulnerable and 500 million people are at extreme risk. Developing countries bear 90 per cent of the climate change burden, 98 per cent of the seriously affected and 99 per cent of all deaths from weather-related disasters. It further argues that climate change exacerbates existing inequalities faced by vulnerable groups, especially women, children and the elderly. Women and children in particular are disproportionately represented among people displaced by extreme weather events and other climate shocks.

The UNDP also highlights the message that climate change will undermine efforts to combat poverty, and that high levels of poverty and low levels of human development limit the capacity of poor households to manage climate risks. Ironically, strategies for coping with climate risks can reinforce deprivation – for example, producers in drought prone areas may well forego the production of crops that could raise their incomes in order to minimise risks; when climate disasters strike, the poor are often forced to sell productive assets to maintain their consumption; falls in income may lead to cuts in expenditure on health and education, trapping vulnerable households in low human development traps (UNDP, 2008: 16).

Economic, social and ecological processes interact with one another and shape the opportunities for economic development in all respects. The UNDP concludes (2008: 27) that 'Combating climate change demands that we place ecological imperatives at the heart of economics. . . . that, with the right reforms, it is not too late to cut green-house gas emissions to sustainable levels without sacrificing economic growth: that rising prosperity and climate security are not conflicting objectives.'

References

Adams, R. 2003. *Economic Growth, Inequality and Poverty: Findings From a New Data Set.* World Bank Working Paper Number 2972. Washington, DC: World Bank.

Adelman, I. 2001. Fallacies in Development Theory and their Implications for Policy. In Meier, G. and Stiglitz, J. (eds) *Frontiers of Development Economics: The Future in Perspective.* New York: Oxford University Press for the World Bank, 103–48.

Adelman, I. and Morris, C. 1973. *Economic Growth and Social Equity in Developing Countries.* Stanford, CA: Stanford University Press.

Agénor, P-R. and Montiel, P.J. 1999. *Development Macroeconomics* (2nd edn). Princeton NJ: Princeton University Press.

Aghion, P., Caroli, E. and Garcia-Peñalosa, C. 1999. Inequality and Economic Growth: The Perspective of the New Growth Theories. *Journal of Economic Literature.* 37 (4): 1615–60.

Ahluwalia, M. 1976. Inequality, Poverty and Development. *Journal of Development Economics.* 3: 307–42.

Ahluwalia, M., Carter, N. and Chenery, H. 1979. Growth and Poverty in Developing Countries. *Journal of Development Economics.* 6: 299–341.

Alesina, A. and Perotti, R. 1996. Income Distribution, Political Instability and Investment. *European Economic Review.* 40: 1203–28.

Alesina, A. and Rodrik, D. 1994. Distributive Policies and Economic Growth. *Quarterly Journal of Economics.* 109: 465–90.

Alkire, S. 2002. *Valuing Freedoms: Sen's Capability Approach and Poverty Reduction.* New York: Oxford University Press.

Allen, T. and Thomas, A. (eds) 2000. *Poverty and Development into the 21st Century.* Oxford: Oxford University Press in association with the Open University.

Amann, A., Aslanidis, N., Nixson, F. and Walters, B. 2006. Economic Growth and Poverty Alleviation: A Reconsideration of Dollar and Kraay. *European Journal of Development Research.* 18 (1): 22–44.

Amsden, A. 1992. *Asia's Next Giant: South Korea and Late Industrialization.* New York: Oxford University Press.

Anand, S. and Kanbur, R. 1993a. The Kuznets Process and the Inequality-Development Relationship. *Journal of Development Economics*. 40: 25–52.

Anand, S. and Kanbur, R. 1993b. Inequality and Development: a Critique. *Journal of Development Economics*. 41: 19–43.

Anand, S. and Sen, A. 2000. Human Development and Economic Sustainability. *World Development*. 28 (12): 2029–49.

Aryeetey, E., Harrigan, J. and Nissanke, M. (eds) 2000. *Economic Reforms in Ghana*. Oxford: James Currey.

Athukorala, P. 1993. Manufactured Exports from Developing Countries and Their Terms of Trade: A Reexamination of the Sarkar-Singer results. *World Development*. 21 (10) October: 1607–13.

Atkinson, A. 1970. On the Measurement of Inequality. *Journal of Economic Theory*. 2: 244–63.

Atkinson, A. 1983. *The Economics of Inequality*. Oxford: Clarendon Press.

Atkinson, A. 1987. On the Measurement of Poverty. *Econometrica*. 55: 749–64.

Baah-Nuakoh, A. 1997. *Studies on the Ghanaian Economy*. Accra: Ghana Universities Press.

Balance, R., Ansari, J. and Singer, H. 1982. *The International Economy and Industrial Development: Trade and Investment in the Third World*. Brighton: Wheatsheaf.

Balassa, B. 1989. Outward Orientation. In Chenery, H. and Srinivasan, T. (eds) *Handbook of Development Economics* Vol. 2. Amsterdam: Elsevier, 1645–90.

Baldwin, R.E. 1969. The Case Against Infant-Industry Tariff Protection. *Journal of Political Economy*. 77 (3) May–June: 295–305.

Banerjee, A. and Duflo, E. 2003. Inequality and Growth: What Can the Data Say? *Journal of Economic Growth*. 8 (3): 67–299.

Baran, P. 1957. *The Political Economy of Growth*. New York: Monthly Review Press.

Bardhan, P. 1988. Alternative Approaches to Development Economics. In Chenery, H. and Srinivasan, T. (eds) *Handbook of Development Economics* Vol. 1. Amsterdam: Elsevier, 39–72.

Bardhan, P. and Ray, I. 2006. *Methodological Approaches in Economics and Anthropology*. Q-Squared Working Paper Number 17. Centre for International Studies, University of Toronto. Downloaded 16 March 2009 from www.q-squared.ca.

Barro, R. 2000. Inequality and Growth in a Panel of Countries. *Journal of Economic Growth*. 5: 5–32.

Baster, N. 1972. Development Indicators: An Introduction. *Journal of Development Studies*. 8 (2): 1–20.

Baster, N. 1979. Models and Indicators. In Cole, S. and Lucas, H. (eds) *Models, Planning and Basic Needs*. Oxford: Pergamon, 99–103.

Besley, T. and Cord, L. (eds) 2006. *Delivering on the Promise of Pro-poor Growth: Insights and Lessons from Country Experiences*. Houndsmills, Basingstoke: Palgrave Macmillan.

Bhagwati, J. 2002. *The Wind of the Hundred Days: How Washington Mismanaged Globalization*. Cambridge, MA: MIT Press.

Bhalla, S. 2002. *Imagine There's No Country: Poverty, Inequality, and Growth in the Era of Globalization*. Washington, DC: Institute for International Economics.

Bigsten, A. 2003. *Prospects for 'Pro-Poor Growth' in Africa*. Paper prepared for United Nations University World Institute of Development Economics (UNU/WIDER) Conference, 'Inequality, Poverty and Human Well-being', Helsinki, 30–31 May.

Binswanger, H. and Lutz, E. 2003. Agricultural Trade Barriers, Trade Negotiations and the Interests of Developing Countries. In Toye, J. (ed.) *Trade and Development: Directions for the 21st Century*. Cheltenham: Edward Elgar, 151–68.

Birdsall, N. and Londõno, J. 1997. Asset Inequality Matters: An Assessment of the World Bank's Approach to Poverty Reduction. *American Economic Review*. 87(2): 32–7.

Birdsall, N., Ross, D. and Sabot, R. 1995 Inequality and Growth Reconsidered: Lessons from East Asia. *World Bank Economic Review*. 9: 477–508.

Bourguignon, F. 1979. Decomposable Income Inequality Measures. *Econometrica*. 47: 901–20.

Bourguignon, F. 2003. The Growth Elasticity of Poverty Reduction: Explaining Heterogeneity Across Countries and Time Periods. In Eicher, T. and Turnovsky, S. (eds) *Inequality and Growth: Theory and Policy Implications*. Cambridge, MA: MIT Press.

Bruno, M., Ravallion, M. and Squire, L. 1998. Equity and Growth in Developing Countries: Old and New Perspectives on the Policy Issues. In Tanzi, V. and Chu, K. (eds) *Income Distribution and High Quality Growth*. Cambridge, MA: MIT Press.

Bruton, H. 1989. Import Substitution. In Chenery, H. and Srinivasan, T. (eds) *Handbook of Development Economics* Vol. 2. Amsterdam: Elsevier, 1637–44.

Burnside, C. and Dollar, D. 2000. Aid, Policies and Growth. *American Economic Review*. 90 (4): 847–68.

Cardoso, F.H. and Faletto, E. 1979. *Dependency and Development in Latin America*. Berkeley: University of California Press.

Cernat, L., Laird, S. and Turrini, A. 2002. *Back to Basics: Market Access Issues in the Doha Agenda*. Trade Analysis Branch, Division on International Trade in Goods and Services, and Commodities. Geneva: United Nations Conference on Trade and Development.

Chambers, R. 1983. *Rural Development: Putting the First Last*. London: Intermediate Technology Development Group.

Chambers, R. 1997. *Whose Reality Counts? Putting the First Last*. London: Intermediate Technology Development Group.

Chambers, R. 2006. Poverty Unperceived: Traps, Biases and Agenda. IDS Working Paper 270. Brighton, UK: Institute of Development Studies.

Chambers, R. and Conway, G. 1992. Sustainable Rural Livelihoods: Practical Concepts for the 21st Century. IDS Discussion Paper 296. Brighton, UK: Institute of Development Studies.

Chang, H.-J. 2002. *Kicking Away the Ladder*. London: Anthem Press.

Chang, H.-J. (ed.) 2003. *Rethinking Development Economics*. London: Anthem Press.

Chang, H.-J. 2003. Trade and Industrial Policy Issues. In Chang, H.-J. (ed.) *Rethinking Development Economics*. London: Anthem Press, 257–76.

Chen, D.H.C., Ranaweera, T. and Storozhuk, A. 2004. *The RMSM-S+P: A Minimal Poverty Module for the RMSM-X*. Washington: World Bank – Policy Research Working Paper WPS3304.

Chenery, H.B. 1960. Patterns of Industrial Growth. *American Economic Review*. 50 (4) September: 624–54.

Chenery, H.B. 1979. *Structural Change and Development Policy*. Oxford: Oxford University Press for the World Bank.

Chenery, H.B. and Syrquin, M. 1975. *Patterns of Development 1950–1970*. Oxford: Oxford University Press for the World Bank.

Chenery, H.B. and Taylor, L. 1968. Development Patterns: Among Countries and Over Time. *Review of Economics and Statistics*. 50 (4) November: 391–416.

Chenery, H.B., Robinson, S. and Syrquin, M. 1986. *Industrialisation and Growth: A Comparative Study*. Oxford: Oxford University Press for the World Bank.

Chenery, H., Ahluwalia, C., Bell, J., Duloy, J. and Jolly, R. 1974. *Redistribution with Growth*. Oxford: Oxford University Press for the World Bank.

Chhibber, A. and Leechor, C. 1995. From Adjustment to Growth in Sub-Saharan Africa: The Lessons of East Asian Experience Applied to Ghana. *Journal of African Economies*. 4 (1) May: 83–114.

Clarke, G. 1995. More Evidence on Income Distribution and Growth. *Journal of Development Economics*. 47 (2): 403–27.

Clarke, R. and Kirkpatrick, C. 1992. Trade Policy Reform: Recent Evidence from Theory and Practice. In Adhikari, R., Kirkpatrick, C. and Weiss, J. (eds) *Industrial and Trade Policy Reform in Developing Countries*. Manchester: Manchester University Press, 56–73.

Clunies-Ross, A., Forsyth, D. and Huq, M. 2009. *Development Economics*. London: McGraw-Hill.

Collier, P. 2000. Africa's Comparative Advantage. In Jalilian, H., Tribe, M. and Weiss, J. (eds) *Industrial Development and Policy in Africa*. Cheltenham: Edward Elgar, 11–21.

Colman, D. 2007. The Common Agricultural Policy. In Artis, M. and Nixson, F. (eds) *The Economics of the European Union: Policy and Analysis* (4th edn). Oxford: Oxford University Press, 77–104.

Colman, D. and Nixson, F. 1994. *Economics of Change in Less Developed Countries* (3rd edn). London: Harvester Wheatsheaf.

Commission for Africa. 2005. *Our Common Interest: Report of the Commission for Africa*. London: Commission for Africa.

Commission of the European Communities (EU) 2001. *European Competitiveness Report 2001*. Brussels: Directorate General Enterprise.

Commission on Growth and Development. 2008. *The Growth Report: Strategies for Sustained Growth and Inclusive Development*. Washington, DC: World Bank (for the Commission on Growth and Development) – downloadable from the Commission on Growth and Development's website.

Convery, F. 1995. Applying Environmental Economics in Africa. World Bank Technical Paper 277, Africa Technical Series. Washington, DC: The World Bank.

Convery, F. and Tutu, K. 1991. *Evaluating the Costs of Environmental Degradation in Ghana*. Accra: Environmental Protection Council.

Cook, P. and Kirkpatrick, C. 2003. Assessing the Impact of Privatization in Developing Countries. In Parker, D. and Saal, D. (eds) *International Handbook on Privatisation*. Cheltenham: Edward Elgar, 209–19.

Cord, L., Lopez, H. and Page, J. 2003. 'When I Use the Word . . .' Pro-Poor Growth and Poverty Reduction. Mimeograph. Washington, DC: World Bank.

Cornia, G.A. (ed.) 2004. *Inequality, Growth and Poverty in an Era of Liberalization and Globalization*. Oxford: Oxford University Press.

Cosgel, M.M. 2006. Conversations Between Anthropologists and Economists. Q-Squared Working Paper Number 18. Centre for International Studies, University of Toronto. Downloaded 16 March 2009 from www.q-squared.ca.

Cypher, J. and Dietz, J. 2004. *The Process of Economic Development* (2nd edn). London: Routledge.

Cypher, J. and Dietz, J. 2008. *The Process of Economic Development* (3rd edn). London: Routledge.

Dagdeviren, H., Van der Hoeven, R. and Weeks, J. 2002. Poverty Reduction with Growth and Redistribution. *Development and Change.* 33(3): 383–413.

Dasgupta, P. 2007. *Economics: A Very Short Introduction*. Oxford: Oxford University Press.

Datt, G. and Ravallion, M. 1992. Growth and Redistribution Components of Changes in Poverty Measures: a Decomposition with Applications to Brazil and India in the 1980s. *Journal of Development Economics.* 38: 275–95.

Deane, P. 1979. *The First Industrial Revolution* (2nd edn). Cambridge: Cambridge University Press.

Deininger, K. and Squire, L. 1996. Measuring Inequality: a New Data Base. *World Bank Economic Review.* 10(3): 565–91.

Deininger, K. and Squire, L. 1998. New Ways of Looking at Old Issues: Inequality and Growth. *Journal of Development Economics.* 57(2): 259–87.

Deutsch, J. and Silber, J. 2004. Measuring the Impact of Various Income Sources on the Link between Inequality and Development. *Review of Development Economics.* 8(1): 110–27.

DFID (Department for International Development) 2005. *Pro-Poor Growth in the 1990s: Lessons and Insights from 14 Countries.* London: Department for International Development.

DFID 2006. *Development Works: 52 Weeks a Year.* London: Department for International Development.

DFID. 2009. *Glossary.* Accessed on 4 June 2009 from www.dfid.gov.uk/About-DFID/Glossary/.

Dicken, P. 2007. *Global Shift: Mapping the Changing Contours of the World Economy* (5th edn). London: Sage.

Dollar, D. and Kraay, A. 2002. Growth is Good for the Poor. *Journal of Economic Growth.* 7 (3): 195–225.

Dollar, D. and Kraay, A. 2004. Growth is Good for the Poor. In Shorrocks, A. and Van Der Hoeven, R. (eds) *Growth, Inequality, and Poverty: Prospects for Pro-Poor Economic Development.* Oxford: Oxford University Press, 62–80.

Domar, E. 1946. Capital Expansion, Rate of Growth, and Employment. *Econometrica.* 14 (2): 137–47.

Dornbusch, R. and Helmers, F.L.C.H. (eds) 1988. *The Open Economy: Tools for Policymakers in Developing Countries.* New York: Oxford University Press for the World Bank.

Dornbusch, R., Fischer, S. and Startz, R. 2004. *Macroeconomics* (9th edn). New York: McGraw-Hill.

Dos Santos, T. 1973. The Crisis of Development Theory and the Problem of Dependence in Latin America. In Bernstein, H. (ed.) *Underdevelopment and Development.* Harmondsworth: Penguin.

Doyal, L. and Gough, I. 1991. *A Theory of Human Need.* Basingstoke: Macmillan.

Durlauf, S. and Blume, L. (eds) 2008. *The New Palgrave Dictionary of Economics* (2nd edn). Houndmills, Basingstoke: Palgrave Macmillan.

Dutt, A. 1988. Monopoly Power and Uneven Development: Baran Revisited. *Journal of Development Studies.* 24 (2) January: 161–76.

Easterly, W. 1999a. The Ghost of Financing Gap: Testing the Growth Model used in the International Financial Institutions. *Journal of Development Economics.* 60 (2): 423–38.

Easterly, W. 1999b. *Life During Growth*. Mimeograph. Washington, DC: World Bank.

Easterly, W. 2002. Inequality Does Cause Underdevelopment. Centre for Global Development Working Paper Number 1. Washington, DC: Centre for Global Development.

Easterly, W. and Rebelo, S. 1993. Fiscal Policy and Economic Growth: An Empirical Investigation. *Journal of Monetary Economics.* 32(3): 417–58.

Easterly, W., Kremer, M., Pritchett, L. and Summers, L. 1993. Good Policy or Good Luck? Country Growth Performance and Temporary Shocks. *Journal of Monetary Economics.* 32(3): 459–83.

Eastwood, R. and Lipton, M. 2001. Pro-Poor Growth and Pro-Poor Growth Poverty Reduction. Paper presented at 'Asia and Pacific Forum on Poverty: Reforming policies and institutions for poverty reduction', Asian Development Bank: Manila.

Economics Network, The. 2009. *The Economics Network*. Resources available from: www.economicsnetwork.ac.uk.

Economist, The. 2009. *The Economics A–Z*. Accessible from: www. economist.com/research/economics/.

Edwards, C. 1985. *The Fragmented World: Competing Perspectives on Trade, Money and Crisis*. London: Methuen.

Ekins, P. and Max-Neef, M. (eds) 1992. *Real Life Economics*. London: Routledge.

ELDIS 2009. *ELDIS*. Accessed on 4 June 2009 from: www.eldis.org/.

Epaulard, A. 2003. Macroeconomic Performance and Poverty Reduction. International Monetary Fund (IMF) Working Paper Number 72. Washington, DC: IMF.

Esteva, G. 1992. Development. In Sachs, W. (ed.). *The Development Dictionary*. London: Zed Books, 6–25.

European Commission. 2001. Theories of Economic Growth. In *European Competitiveness Report 2001*. Annex II.1 Directorate General Enterprise. Luxembourg: Office for Official Publications of the European Communities (OOPEC), 29–32.

European Union. 2009. *Regional Policy: Social Cohesion – Poverty*. Accessed on 16 November 2009 from: http://ec.europa.eu/regional_policy/sources/docoffic/official/reports/p122_en.htm.

Evans, D. 1989. Alternative Perspectives on Trade and Development. In Chenery, H. and Srinivasan, T. (eds) *Handbook of Development Economics* Vol. 2. Amsterdam: Elsevier, 1241–304.

Fields, G. 2001. *Distribution and Development: A New Look at the Developing World*. Cambridge, MA: MIT Press.

Fine, B. 2002. Economics Imperialism and the New Development Economics as Kuhnian Paradigm Shift? *World Development*. 30(12) December: 2057–70.

Fine, B. 2006a. New Growth Theory: More Problem than Solution. In Jomo, K.S. and Fine, B. (eds) *The New Development Economics: After the Washington Consensus*. London: Zed Books, 68–86.

Fine, B. 2006b. The Developmental State and the Political Economy of Development. In Jomo, K.S. and Fine, B. (eds) *The New Development Economics: After the Washington Consensus*. London: Zed Books, 101–22.

Fitzgerald, E.V.K., Cubero-Brealey, R. and Lehmann, A. 1998. The Development Implications of the Multilateral Agreement on Investment. A report commissioned by the Department for International Development from the Finance and Trade Policy Research Centre, Queen Elizabeth House, University of Oxford: Oxford. Accessed 26 June 1998 from: www.dfid.gov.uk/multinv.htm.

Forbes, K. 2000. A Reassessment of the Relationship between Inequality and Growth. *American Economic Review*. 40(4): 869–87.

Fortin, C. 1988. International Commodities and Third World Development: Trends and Prospects. In Stokke, O. (ed.) *Trade and Development: Experiences and Challenges*. Tilburg: European Association of Development Training and Research Institutes, 59–76.

Foster, J. and Székely, M. 2002. Is Economic Growth Good for the Poor? Tracking Low Incomes Using General Means. Paper presented at United Nations University World Institute for Development Economics Research (UNU/WIDER) conference, 'Growth and poverty', Helsinki, 25–6 May.

Foster, J., Greer, J. and Thorbecke, E. 1984. A Class of Decomposable Poverty Measures. *Econometrica*. 52: 761–66.

Frank, A.G. 1969. *Latin America: Underdevelopment or Revolution*. New York: Monthly Review Press.

Gallup, J., Radelet, S. and Warner, A. 1999. Economic Growth and the Income of the Poor. Consulting Assistance on Economic Reform II Discussion Paper Number 36. Harvard, MA: Harvard Institute of International Development.

Ghosh, D. 2007. The Metamorphosis of Lewis's Dual Economy Model. *Journal of Economic Methodology*. 14 (1) March: 5–25.

Global Humanitarian Forum. 2009. Human Impact Report: Climate Change – The Anatomy of A Silent Crisis. Geneva: Global Humanitarian Forum – accessed on 5 June 2009 from: www.ghf-ge.org/programmes/human_impact_report.

Google. 2009. *Google*. Accessed on 4 June 2009 from: www.google.co.uk/.

Google Books. 2009. *Google Books*. Accessed on 4 June 2009 from: books.google.co.uk/.

Google Scholar. 2009. *Google Scholar*. Accessed on 4 June 2009 from: scholar.google.co.uk/.

Gore, C. 2000. The Rise and Fall of the Washington Consensus as Paradigm for Developing Countries. *World Development*. 28(5): 789–804.

Greenaway, D. 1991. New Trade Theories and Developing Countries. In Balasubramanyam, V.N. and Lall, S. (eds) *Current Issues in Development Economics*. London: Macmillan, 156–70.

Greenaway, D. and Milner, C. 1987. Trade Theory and the Less Developed Countries. In Gemmell, N. (ed.) *Surveys in Development Economics*. Oxford: Basil Blackwell, 11–55.

Greenaway, D., Morgan, W. and Wright, P. 1998. Trade Reform, Adjustment and Growth: What Does the Evidence Tell Us? *Economic Journal*. CVIII (450) September: 1547–61.

Grimm, M., Klasen, S. and McKay, A. 2007. *Determinants of Pro-Poor Growth: Analytical Issues and Findings from Country Cases*. Houndsmills, Basingstoke: Palgrave Macmillan.

Group of 77 and China. 2001. *Declaration by the Group of 77 and China on the Fourth WTO Ministerial Conference at Doha, Qatar*. Group of 77, Geneva. Downloaded 26 January 2005 from the G77 website: www.g77.org/Docs/Doha.htm.

Gylfason, T. 2008. Dutch Disease. In Durlauf, S. and Blume, L. (eds) *The New Palgrave Dictionary of Economics* (2nd edn). Houndmills, Basingstoke: Palgrave Macmillan.

Harriss, J. 2002. The Case for Cross-disciplinary Approaches in International Development. *World Development*. 30 (3) March: 487–96.

Harrod, R.F. 1939. An Essay in Dynamic Theory. *Economic Journal*. 49 (193): 14–33.

Hayter, T. 1971. *Aid as Imperialism*. Harmondsworth: Penguin.

HM Treasury 2008. *The Green Book: Appraisal and Evaluation in Central Government*. Accessed on 19 June 2008 from the Treasury website: www.hm-treasury.gov.uk/economic_data_and_tools/greenbook/data_greenbook_index.cfm.

Heston, A., Summers, R. and Aten, B. 2006. *Penn World Table Version 6.2*, Center for International Comparisons of Production, Income and Prices at the University of Pennsylvania, September.

Hicks, N. and Streeten, P. 1979. Indicators of Development: The Search for a Basic Needs Yardstick. *World Development*. 7: 567–80.

Higgins, B.H. 1968. *Economic Development: Principles, Problems and Policies* (2nd edn). London: Constable.

Hirschman, A.O. 1958. *The Strategy of Economic Development*. New Haven: Yale University Press.

Hirst, P. and Thompson, G. 1996. *Globalization in Question*. Cambridge: Polity Press.

Hirst, P. and Thompson, G. 2003. The Future of Globalisation. In Michie, J. (ed.) *The Handbook of Globalisation*. Cheltenham: Edward Elgar, 17–36.

Hoekman, B., Mattoo, A. and English, P. (eds) 2002. *Development, Trade, and the WTO: A Handbook*. Washington: World Bank.

Hulme, D., Moore, K. and Shepherd, A. 2001. Chronic Poverty: Meanings and Analytical Frameworks. Chronic Poverty Research Centre (CPRC) Working Paper Number 2. Manchester and London: CPRC.

Huq, M. 1989. *The Economy of Ghana: The First 25 Years since Independence*. Houndmills, Basingstoke: Macmillan.

Husain, I. 1994. Results of Adjustment in Africa: Selected Cases. *Finance and Development*. 31 (2): 6–9.

id21 2009. *id21*. Accessed on 4 June 2009 from: www.id21.org/.

IDD 2006. *Joint Evaluation of General Budget Support 1994–2004*; Birmingham: University of Birmingham International Development Department. Accessed 6 May 2009 from: www.oecd.org/dataoecd/.

ILO 1976. *Employment, Growth and Basic Needs: A One-World Problem*. Geneva: International Labour Office.

ILO 1977. *Meeting Basic Needs: Strategies for Eradicating Mass Poverty and Unemployment*. Geneva: International Labour Office.

ILO 2004. *A Fair Globalization: Creating Equal Opportunities for All*. Geneva: International Labour Office for the World Commission on the Social Dimensions of Globalization.

Imbs, J. and Wacziarg, R. 2003. Stages of Diversification. *American Economic Review*. 93 (1) March: 63–86.

IMF 1997. *World Economic Outlook*. Washington: International Monetary Fund.

IMF 2009. *Poverty Reduction Strategy Papers (PRSP)*. Accessed on 16 May 2009 from the International Monetary Fund's website at: www.imf.org/external/np/prsp/prsp.asp.

Ingham, B. 1995. *Economics and Development*. London: McGraw-Hill.

Intergovernmental Panel on Climate Change 2007. *Fourth Assessment Report: Synthesis Report*. Geneva: Intergovernmental Panel on Climate Change. Accessible from the IPCC website: http://www.ipcc.ch/publications_and_data/publications_ipcc_fourth_assessment_report_synthesis_report.htm.

IOM 2009. *Moldova: Migration and Development*. Geneva: International Organization for Migration. Accessed on 2 December 2009 from: www.iom.md/migration&development.html.

Jackson, C. 2002. Disciplining Gender? *World Development*. 30 (3) March: 497–509.

Jalilian, H. and Kirkpatrick, C. 2005. Does Financial Development Contribute to Poverty Reduction? *Journal of Development Studies*. 41(4): 636–56.

Jomo K.S. and Fine, B. (eds) 2006 *The New Development Economics: After the Washington Consensus*. London: Zed Books, 68–86.

Kalwij, A. and Verschoor, A. 2004. How Good is Growth for the Poor? The Role of Initial Income Distribution in Regional Diversity in Poverty Trends. Paper prepared for United Nations University World Institute of Development Economics Research (UNU/WIDER) Conference, 'Sharing global prosperity', Helsinki, 6–7 September.

Kanbur, R. 2001. Economic Policy, Distribution and Poverty: the Nature of Disagreements. *World Development.* 29(6):1083–94.

Kanbur, R. 2002. Economics, Social Science and Development. *World Development.* 30 (3) March: 477–86.

Kanbur, R. 2004. *Growth, Inequality and Poverty: Some Hard Questions.* Mimeograph. Ithaca, NY: Cornell University.

Kanbur, R. and Lustig, N. 1999. Why is Inequality Back on the Agenda? Paper prepared for the Annual Bank Conference on Development Economics (ABCDE), Washington, DC: World Bank, 28–30 April.

Kanbur, R. and Squire, L. 2001. The Evolution of Thinking About Poverty: Exploring the Contradictions. In Meier, G. and Stiglitz, J. (eds) *Frontiers of Development Economics.* Oxford: Oxford University Press, 183–226.

Kaplinsky, R. and Messner, D. 2008. Introduction: The Impact of Asian Drivers on the Developing World. *World Development.* 36 (2) February: 197–209.

Kay, C. 1989. *Latin American Theories of Development and Underdevelopment.* London: Routledge.

Kenny, C. and Williams, D. 2001. What Do We Know About Economic Growth? Or, Why Don't We Know Very Much? *World Development.* 29 (1): 1–22.

Keynes, J.M. 1936. *The General Theory of Employment, Money and Interest.* London: Macmillan.

Killick, A., Gunatilaka, R. and Marr, R. 1998. *Aid and the Political Economy of Policy Change.* London: Routledge.

Kindleberger, C.P. 1958. *Economic Development.* New York: McGraw-Hill.

Kingdon, G. and Knight, J. 2004. Do People Mean What They Say? Implications for Subjective Survey Data. GPRG Working Paper 3. Oxford: Global Poverty Research Group.

Kirkpatrick, C. 1987. Trade Policy and Industrialization in LDCs. In Gemmell, N. (ed.) *Surveys in Development Economics.* Oxford: Basil Blackwell, 56–89.

Kirkpatrick, C. and Barrientos, A. 2004. The Lewis Model After 50 Years. *Manchester School.* 72 (6) December: 679–90.

Klasen, S. 2005. *Economic Growth and Poverty Reduction: Measurement and Policy Issues.* OECD Development Centre Working Paper Number 246. Paris: OECD.

Klugman, J. (ed.) 2002. *PRSP Sourcebook.* Washington: World Bank – see particularly Chapter 6 on Public Spending by A. Fozzard, M. Holmes, J. Klugman, and K. Withers.

Knowles, S. 2005. Inequality and Economic Growth: the Empirical Relationship Reconsidered in Light of Comparable Data. *Journal of Development Studies.* 41(1): 135–59.

Kraay, A. 2004. When is Growth Pro-Poor? Cross-Country Evidence. World Bank Working Paper Number 3225. Washington, DC: World Bank.

Kravis, I. 1970. Trade as a Handmaiden of Growth: Similarities Between the Nineteenth and Twentieth Centuries. *Economic Journal*. 80 (320): 850–72.

Krueger, A. 1997. Trade Policy and Economic Development: How We Learn. *American Economic Review*. 87 (1) March: 1–22.

Krugman, P. 1981. Trade, Accumulation, and Uneven Development. *Journal of Development Economics*. 8 (2) April: 149–61.

Krugman, P. 1997. The Fall and Rise of Development Economics. In Krugman P. (ed.) *Development, Geography, and Economic Theory*. Cambridge, MA: The MIT Press.

Krugman, P. 2008. *The Return of Depression Economics and the Crisis of 2008*. Harmondsworth: Penguin Books.

Kuznets, S. 1955. Economic Growth and Income Inequality. *American Economic Review*. 45(1): 1–28.

Kuznets, S. 1956. Quantitative Aspects of the Economic Growth of Nations. *Economic Development and Cultural Change*. 5 (1): 5–94.

Kuznets, S. 1963. Quantitative Aspects of the Economic Growth of Nations: VIII, Distribution and Income by Size. *Economic Development and Cultural Change*. 11 (2): 1–80.

Kuznets, S. 1971. *Economic Growth of Nations: Total Output and Production Structure*. Cambridge, MA: The Belknap Press of Harvard University Press.

Kuznets, S. 1973. Modern Economic Growth: Findings and Reflections. *American Economic Review*. 63 (3): 247–58.

Lall, S. 2000. The Technological Structure and Performance of Developing Country Manufactured Exports, 1985–98. *Oxford Development Studies*. 28 (3) October: 337–69.

Lall, S., Weiss, J. and Zhang, J. 2006. The 'Sophistication' of Exports: A New Trade Measure. *World Development*. 34 (2): 222–37.

Lee, F. 2008. Heterodox Economics. In Durlauf, S. and Blume, L. (eds) *The New Palgrave Dictionary of Economics* (2nd edn). Houndmills, Basingstoke: Palgrave Macmillan.

Leeson, P. 1988. Development Economics and the Study of Development. In Leeson, P. and Minogue, M. (eds) *Perspectives on Development: Cross-disciplinary Themes in Development*. Manchester: Manchester University Press, 1–55.

Leeson, P. and Minogue, M. (eds) 1988. *Perspectives on Development: Cross-disciplinary Themes in Development*. Manchester: Manchester University Press.

Leeson, P. and Nixson, F. 1988. Development Economics and the State. In Leeson, P. and Minogue, M. (eds) *Perspectives on Development: Cross-disciplinary Themes in Development*. Manchester: Manchester University Press, 56–88.

Leeson, P. and Nixson, F. 2004. Development Economics in the Department of Economics at the University of Manchester. *Journal of Economic Studies*. 31 (1): 6–24.

Leibenstein, H. 1957. *Economic Backwardness and Economic Growth: Studies in the Theory of Economic Development.* New York: Wiley.

Lewis, W.A. 1953. *Report on Industrialisation and the Gold Coast.* Accra: Government Printing Department.

Lewis, W.A. 1954. Economic Development with Unlimited Supplies of Labour. *Manchester School.* 22(2) May: 139–91. Reprinted in Agarwala, A.N. and Singh, S.P. (eds) 1963. *The Economics of Underdevelopment.* New York: Oxford University Press, 400–49.

Lewis, W.A. 1955. *The Theory of Economic Growth.* London: George Allen and Unwin.

Li, H., Squire, L. and Zou, H. 1998. Explaining International and Intertemporal Variations in Income Inequality. *Economic Journal.* 108: 26–43.

Lipsey, R. and Lancaster, K. 1956. The General Theory of Second Best. *Review of Economic Studies.* 24 (1): 11–32.

List, J. and Gallet, C. 1999. The Kuznets Curve: What Happens after the Inverted-U? *Review of Development Economics.* 3: 200–6.

Little, I.M.D., Scitovsky, T. and Scott, M.Fg. 1970. *Industry and Trade in Some Developing Countries.* Oxford: Oxford University Press for the OECD.

Lopez, J. 2005. *Pro-poor Growth: A Review of What We Know (and What We Don't).* Mimeograph. Washington, DC: World Bank.

McClelland, D.C. 1967. *The Achieving Society.* London: The Free Press.

McGillivray, M. 1991. Redundant Composite Development Indicator. *World Development.* 19 (10): 1461–69.

McGillivray, M. 2006. *Inequality, Poverty and Wellbeing.* Houndsmills, Basingstoke: Palgrave Macmillan.

McGillivray, M., Feeny, S., Hermes, N. and Lensink, R. 2006. Controversies over the Impact of Development Aid: It Works; It Doesn't; It Can, but that Depends. *Journal of International Development.* 18 (7): 1031–50.

McGranahan, D., Pizarro, E. and Richard, C. 1985. *Measurement and Analysis of Socio-Economic Development: An Enquiry into International Indicators of Development and Quantitative Interrelations of Social and Economic Components of Development.* Geneva: United Nations Research Institute for Social Development (UNRISD).

McGregor, A. 2007. Researching Wellbeing: From Concepts to Methodology. In Gough, I. and McGregor, A. (eds) *Wellbeing in Developing Countries.* Cambridge: Cambridge University Press, 316–49.

McKay, A. and Lawson, D. 2003. Assessing the Extent and Nature of Chronic Poverty in Low Income Countries: Issues and Evidence. *World Development.* 31 (3): 1–20.

McKay, A. and Sumner, A. 2008. Economic Growth, Inequality and Poverty Reduction: Does Pro-Poor Growth Matter? IDS InFocus 3. March. Brighton: Institute of Development Studies.

McKinnon, R. 1973. *Money and Capital in Economic Development.* Washington: Brookings Institution.

Maizels, A. (ed.) 1987. Primary Commodities in the World Economy: Problems and Policies. *World Development.* 15 (5) May: 537–758.

Maizels, A. 2003. Economic Dependence on Commodities. In Toye, J. (ed.) *Trade and Development: Directions for the 21st Century.* Cheltenham: Edward Elgar, 169–84.

Malthus, T. 1970. *An Essay on the Principle of Population and a Summary View of the Principle of Population* (8th edn). Harmondsworth: Penguin Books (originally published 1878).

Mandeville, B. 1970. *The Fable of the Bees.* Harmondsworth: Penguin Books (originally published in French 1714).

Marx, K. 1853. The East India Company – its History and Results, and The Results of the British Rule of India. Reprinted in Fernbach, D. (ed.) 1973. *Karl Marx, Surveys from Exile: Political Writings* Vol. 2. Harmondsworth: Penguin.

Marx, K. 1999. *Capital.* Oxford: Oxford University Press (originally published 1867 to 1895).

Mavrotas, G. and Shorrocks, A. (eds.) 2007. *Advancing Development: Core Themes in Global Economics.* Houndmills, Basingstoke: Palgrave Macmillan.

Maxwell, S. 1998. Comparisons, Convergence and Connections: Development Studies in North and South. *IDS Bulletin.* 29 (1): 20–31.

Maxwell, S. 1999. What Can We Do with a Rights Based Approach? Overseas Development Institute Briefing Paper. London: ODI.

Maxwell, S. 2005. The Washington Consensus is Dead! Long Live the Meta-Narrative! Overseas Development Institute (ODI) Working Paper 243. London: ODI.

Mayer, J. 2002. The Fallacy of Composition: A Review of the Literature. *The World Economy.* 25 (6) June: 875–94.

Mearman, A. 2007. Teaching Heterodox Economics Concepts. In The Economics Network. *The Handbook for Economic Lecturers.* Downloaded on 20 October 2009 from: www.economicsnetwork.ac.uk/handbook/heterodox.

Meier, G.M. 1964. *Leading Issues in Development Economics* (1st edn). New York: Oxford University Press.

Meier, G. 1990. Trade Policy and Development. In Scott, M. and Lal, D. (eds) *Public Policy and Economic Development.* Oxford: Clarendon Press, 155–70.

Meier, G. (ed.) 1995. *Leading Issues in Economic Development* (6th edn). New York: Oxford University Press.

Meier, G. and Rauch, J. (eds) 2005. *Leading Issues in Economic Development* (8th edn). Oxford: Oxford University Press.

Migration Policy Institute. 2007. *Top Ten Countries with the Largest*

Amounts of Remittances. Washington: Migration Policy Institute. Accessed 12 December 2009 from: www.migrationinformation.org/datahub/remittances/TopTen_06.pdf.

Milanovic, B. 2007. *Worlds Apart: Measuring International and Global Inequality.* Princeton and Oxford: Princeton University Press.

Mohan, G., Brown, E., Milward, B. and Zack-Williams, A. 2000. *Structural Adjustment: Theory, Practice and Impacts.* London: Routledge.

Montgomery, K. 2009. The Demographic Transition. Accessed 16 September 2009 from: www.uwmc.uwc.edu/geography/Demotrans/demtran.htm.

Morris, C. 1979. *Measuring the Condition of the World's Poor: The Physical Quality of Life Index.* London: Frank Cass.

Morrissey, O. 2001. Does Aid Increase Growth? *Progress in Development Studies.* 1 (1): 37–50.

Mosley, P. 2004. Severe Poverty and Growth: A Macro-Micro Analysis. Chronic Poverty Research Centre (CPRC) Working Paper, Number 51. Manchester: CPRC.

Mosley, P., Hudson, J. and Verschoor, A. 2004. Aid, Poverty Reduction and the New Conditionality. *Economic Journal.* 114: 214–43.

Mosley, P., Toye, J. and Harrigan, J. 1995. *Aid and Power: The World Bank and Policy-based Lending. Vol. 1 – Analysis and Policy Proposals.* London: Routledge.

Myint, H. 1958. The 'Classical Theory' of International Trade and the Underdeveloped Countries. *Economic Journal.* LXVIII (274) June: 317–37. Reprinted in Myint, H. 1971. *Economic Theory and the Underdeveloped Countries.* New York: Oxford University Press, 118–46.

Myint, H. 1965. Economic Theory and the Underdeveloped Countries. *Journal of Political Economy.* LXXIII (5) October: 477–91. Reprinted in Myint, H. 1971. *Economic Theory and the Underdeveloped Countries.* New York: Oxford University Press, 3–26.

Myint, H. 1967. Economic Theory and Development Policy. *Economica.* 32 (134) May: 117–30. Reprinted in Myint, H. 1971. *Economic Theory and the Underdeveloped Countries.* Oxford: Oxford University Press, 27–46.

Myint, H. 1972. *South East Asia's Economy: Development Policies in the 1970s.* Harmondsworth: Penguin Books.

Myint, H. 1987. The Neoclassical Resurgence in Development Economics: Its Strengths and Limitations. In Meier, G. (ed.) *Pioneers in Development* 2nd series. New York: Oxford University Press for the World Bank, 105–36.

Myrdal, G. 1957. *Economic Theory and Under-developed Regions.* London: Duckworth.

Myrdal, G. 1968. *Asian Drama.* Harmondsworth: Penguin Books, Vol. III, App. 3.II.

Myrdal, G. 1970. *The Challenge of World Poverty: A World Anti-Poverty Programme in Outline.* Harmondsworth: Penguin Books.

Narayan, N., Patel, R., Schafft, K., Rachemacher, A. and Koch-Schulte, S. 1999. *Voices of the Poor: Can Anyone Hear Us?* Washington, DC: World Bank.

Narlikar, A. 2003. *International Trade and Developing Countries: Bargaining Coalitions in the GATT and WTO.* London: Routledge.

National Bureau of Statistics Tanzania. 2002. *Statistical Abstract 2002.* Dar es Salaam: National Bureau of Statistics. Available from: www.nbs.go.tz/.

Newbery, D. 1990. Commodity Price Stabilization. In Scott, M. and Lal, D. (eds) *Public Policy and Economic Development.* Oxford: Clarendon Press, 80–108.

Newbold, P., Pfaffenzeller, S. and Rayner, A. 2005. How Well Are Long-Run Commodity Price Series Characterized By Trend Components? *Journal of International Development.* 17 (4) May: 479–94.

Nishimizu, M. and Page, J. 1988. Total Factor Productivity. In Meier, G., Steel, W. and Carroll, R. (eds) *Industrial Adjustment in Sub-Saharan Africa.* Washington: World Bank Economic Development Institute, 263–67.

Nishimizu, M. and Robinson, S. 1986. Productivity Growth in Manufacturing. In H. Chenery, S. Robinson and M. Syrquin (eds) *Industrialization and Growth: A Comparative Study.* New York: Oxford University Press, 177–206.

Nissanke, M. 2007. Donor–Recipient Relationships in the Aid Effectiveness Debate. In Jerve, A., Shimomura, Y. and Hansen, A. (eds) *Aid Relationships in Asia: Exploring Ownership in Japanese and Nordic Aid.* Basingstoke: Palgrave Macmillan.

Nissanke, M. and Ferrarini, B. 2007. Assessing the Aid Allocation and Debt Sustainability Framework: Working Towards Incentive Compatible Aid Contracts. Research Paper 2007/33. Helsinki: United Nations University – World Institute for Development Economics Research (UNU – WIDER).

Nixson, F.I. 1990. Industrialisation and Structural Change in Developing Countries. *Journal of International Development.* 2 (3) July: 310–33.

Nixson, F. 2002. Economic Growth and Development in an Unequal World. In Atkinson, G.B.J. (ed.) *Developments in Economics: An Annual Review.* 18. London: Causeway Press, 119–35.

Nixson, F. 2006. Rethinking the Political Economy of Development: Back to Basics and Beyond. *Journal of International Development.* 18 (7): 867–981.

Nixson, F. 2007. Aid, Trade and Economic Development: The EU and the Developing World. In Artis, M. and Nixson, F. (eds) *The Economics of the European Union: Policy and Analysis* (4th edn). Oxford: Oxford University Press,: 322–53.

Nixson, F. 2007–8. *Overseas Aid – Is it Working?* Transactions of the Manchester Statistical Society. Manchester: Manchester Statistical Society.

Nixson, F. and Walters, B. (1999) The Asian Crisis: Causes and Consequences. *Manchester School*. 67 (5): 496–523.

Nussbaum, M. 2000. *Women and Human Development: The Capabilities Approach*. Cambridge: Cambridge University Press.

Nye, N., Reddy, S. and Watkins, K. 2002. *Dollar and Kraay on 'Trade, Growth and Poverty': A Critique*. Mimeograph. New York: Columbia University.

Ocampo, J. 1986. New Developments in Trade Theory and LDCs. *Journal of Development Economics*. 22 (1) June: 129–70.

OECD 1996. *Shaping the 21st Century: The Contribution of Development Co-operation*. Paris: Organisation for Economic Co-operation and Development.

OECD 2001. *Development Assistance Committee Poverty Guidelines*. Paris: Organisation for Economic Co-operation and Development.

OECD 2007. *Promoting Pro-Poor Growth: Policy Guidance for Donors*. DAC Guidelines and Reference Series. Paris: Organisation for Economic Co-operation and Development.

OECD 2009. *Political Economy Analysis*. Website of the Organisation for Economic Co-operation and Development. Accessed 6 May 2009 from: www.oecd.org/dac/governance/politicaleconomy.

O'Neill, H. 1997. Globalisation, Competitiveness and Human Security: Challenges for Development Policy and Institutional Change. *European Journal of Development Research*. 9 (1): 7–37.

Oxfam. 2002. *Rigged Rules and Double Standards: Trade, Globalisation, and the Fight against Poverty*. Oxford: Oxfam.

Pack, H. 1989. Industrialization and Trade. In Chenery, H. and Srinivasan, T. (eds) *Handbook of Development Economics* Vol. 1. Amsterdam: Elsevier, 333–80.

Pack, H. 1994. Endogenous Growth Theory: Intellectual Appeal and Empirical Shortcomings. *Journal of Economic Perspectives*. 8 (1): 55–72.

Parker, D. and Saal, D. (eds) 2005. *International Handbook on Privatisation*. Cheltenham: Edward Elgar.

Paukert, F. 1973. Income Distribution at Different Levels of Development. *International Labour Review*. 108: 97–125.

Peacock, A. and Dosser, D. 1958. *The National Income of Tanganyika*. Colonial Research Studies 26. London: Colonial Office.

Perkins, D., Radelet, S., Snodgrass, D., Gillis, M. and Roemer, M. 2001. *Economics of Development* (5th edn). New York: Norton.

Perotti, R. 1996. Growth, Income Distribution and Democracy: What the Data Say. *Journal of Economic Growth*. 1(2): 149–87.

Polak, J.J. 1997. The IMF Monetary Model: A Hardy Perennial. *Finance and Development*. 34 (4): 16–19.

Prebisch R. 1950. *The Economic Development of Latin America and its Principal Problems*. Santiago: United Nations Economic Commission for Latin America.

Quesnay, F. 1973. *The Economical Table (Tableau Economique)*. New York: Gordon Publishing (translated by the Marquis de Mirabeau and reproduced from the original 1766 edition).

Ranis, G. and Fei, J. 1961. A Theory of Economic Development. *American Economic Review*. 51 (4) September: 533–65.

Ravallion, M. 1995. Growth and Poverty: Evidence for Developing Countries in the 1980s. *Economic Letters*. 48: 411–17.

Ravallion, M. 1997. Can High-inequality Developing Countries Escape Absolute Poverty? *Economic Letters*. 56: 51–7.

Ravallion, M. 1998. Does Aggregation Hide the Harmful Effects of Inequality on Growth? *Economic Letters*. 61: 73–7.

Ravallion, M. 2001. Growth, Inequality and Poverty: Looking Behind the Averages. *World Development*. 29(11): 1803–15.

Ravallion, M. 2003. Targeted Transfers in Poor Countries: Revisiting the Trade-Offs and Policy Options. Social Protection Discussion Paper 314. Washington, DC: World Bank.

Ravallion, M. 2004. Measuring Pro-poor Growth: A Primer. World Bank Working Paper 3242. Washington, DC: World Bank.

Ravallion, M. and Chen, S. 1997. What Can New Survey Data Tell Us About Recent Changes in Distribution and Poverty. *World Bank Economic Review*. 11(2): 357–82.

Ravallion, M. and Chen, S. 2003. Measuring Pro-poor Growth. *Economic Letters*. 73: 93–9.

Ravenhill, J. (ed.) 2005a. *Global Political Economy*. Oxford: Oxford University Press.

Ravenhill, J. 2005b. Regionalism. In Ravenhill, J. (ed.) *Global Political Economy*. Oxford: Oxford University Press, 116–47.

Rawlings, L. and Rubio, G. 2005. Evaluating the Impact of Conditional Cash Transfer Programs. *World Bank Research Observer*. 20 (1): 29–55.

Ray, D. 2008. Development Economics. In Durlauf, S. and Blume, L. (eds) *The New Palgrave Dictionary of Economics* (2nd edn). Houndmills, Basingstoke: Palgrave Macmillan. Accessed from *The New Palgrave Dictionary of Economics Online* 4 June 2009 from: www.dictionaryof economics.com.

Reddaway, W.B. 1962. *The Development of the Indian Economy*. London: George Allen and Unwin.

Rehme, G. 2001. *Redistribution of Personal Income, Education and Economic Performance Across Countries*. Mimeograph. Technische Universitat Darmstadt: Darmstadt, Germany.

Reinikka, R. and Collier, P. 2001. *Uganda's Recovery: The Role of Farms, Firms and Government*. Washington, DC: World Bank.

Republic of Ghana. Various issues. *Quarterly Digest of Statistics*. Accra: Central Bureau of Statistics, and Ghana Statistical Service.

Republic of Ghana. 1996. *Measuring Informal Sector Activity in Ghana: Proceedings of a Ghana Statistical Service/Overseas Development Administration Workshop*. Accra: Ghana Statistical Service.

Republic of Uganda. Various years. *Background to the Budget*. Kampala: Ministry of Finance.

Republic of Uganda. 2008. *Statistical Abstract*. Kampala, Uganda Bureau of Statistics. Available from: www.ubos.org/.

Ricardo, D. 1953. *On the Principles of Political Economy and Taxation* (3rd edn) (edited by P. Sraffa with the assistance of M.H. Dobb). Cambridge: Cambridge University Press for the Royal Economic Society.

Riddell, R. 2007. *Does Foreign Aid Really Work?* Oxford: Oxford University Press.

Robeyns, I. 2005. The Capability Approach: a Theoretical Survey. *Journal of Human Development*. 6 (2): 93–117.

Rodriguez, F. and Rodrik, D. 1999. Trade Policy and Economic Growth: A Sceptic's Guide to the Cross-National Evidence. Discussion Paper 2143. London: Centre for Economic Policy Research. Accessed 24 March 2005 from: www.cepr.org/pubs/new-dps/dplist.asp?dpno=2143. Also available from the University of Maryland: www.bsos.umd.edu/econ/papers/rodriguez9901.pdf.

Rodrik, D. 1995. Trade and Industrial Policy Reform. In Behrman, J. and Srinivasan, T. (eds) *Handbook of Development Economics* Vol. 3B. Amsterdam: Elsevier, 2925–82.

Rodrik, D. 2002. After Neoliberalism, What? Paper presented at 'Alternatives to Neoliberalism'. Washington, DC, 23 May. Downloaded from: http://ksghome.harvard.edu/~drodrik.

Rodrik, D. 2007. *One Economics: Many Recipes: Globalization, Institutions and Economic Growth*. Princeton and Oxford: Princeton University Press.

Rodrik, D. 2009. A Plan B for Global Finance. *The Economist*. 12 March.

Roemer, M. and Gugerty, M. 1997. Does Economic Growth Reduce Poverty? Consulting Assistance on Economic Reform II Discussion Paper Number 4. Harvard, MA: Harvard Institute of International Development.

Romer, P.M. 1994. The Origins of Endogenous Growth. *Journal of Economic Perspectives*. 8(1): 3–22.

Rosenberg, N. 2004. *Innovation and Economic Growth*. Paris: Organisation for Economic Co-operation and Development. Accessed 3 March 2009 from: www.oecd.org/dataoecd/55/49/34267902.pdf.

Rosenstein-Rodan, P.N. 1943. Problems of Industrialization in Eastern and South-Eastern Europe. *Economic Journal*. 53 (210/211) June: 202–11. Reprinted in Agarwala, A.N. and Singh, S.P. (eds) 1963. *The Economics of Underdevelopment*. New York: Oxford University Press, 245–55.

Rostow, W.W. 1960a. *The Stages of Economic Growth* (2nd edn). Cambridge: Cambridge University Press.

Rostow, W.W. 1960b. *The Process of Economic Growth* (2nd edn). Oxford: Clarendon Press.

Roxborough, I. 1979. *Theories of Underdevelopment*. London: Macmillan.

Sachs, J.D. and Larrain, B.F. 1992. *Macroeconomics in the Global Economy*. Hemel Hempstead: Harvester Wheatsheaf.

Sachs, J. and Warner, A. 1995. *Economic Reform and the Process of Global Integration*. Washington: Brookings Institution.

Sagar, A. and Najam, A. 1999. Shaping Human Development: Which Way Next? *Third World Quarterly*. 20(4): 743–51.

Salvatore, D. 2004. *International Economics* (8th edn). Chichester: Wiley.

Sarkar, P. and Singer, H. 1993. Manufactured Exports of Developing Countries and their Terms of Trade since 1965: A Reply. *World Development*. 21(10) October: 1617–20.

Sapsford, D. 1985. The Statistical Debate on the Net Barter Terms of Trade Between Primary Commodities and Manufactures: A Comment and Some Additional Evidence. *Economic Journal*. 95(379) September: 781–88.

Sapsford, D. 1988. The Debate over Trends in the Terms of Trade. In Greenaway, D. (ed.) *Economic Development and International Trade*. London: Macmillan, 117–31.

Sapsford, D. and Balasubramanyam, V.N. 1994. The Long-Run Behaviour of the Relative Price of Primary Commodities: Statistical Evidence and Policy Implications. *World Development*. 22 (11) November: 1737–45.

Sapsford, D. and Chen, J.-R. (eds) 1999. Policy Arena – The Prebisch-Singer Thesis: A Thesis for the New Millennium? *Journal of International Development*. 11 (6) September–October: 843–916.

Sapsford, D., Sarkar, P. and Singer, H. 1992. The Prebisch-Singer Terms of Trade Controversy Revisited. *Journal of International Development*. 4 (3) May–June: 315–32.

Sarkar, P. and Singer, H. 1991. Manufactured Exports of Developing Countries and their Terms of Trade since 1965. *World Development*. 19 (4) April: 333–40.

Schumpeter, J. 1961. *The Theory of Economic Development*. New York: Oxford University Press.

Schumpeter, J. 1987. *Capitalism, Socialism and Democracy* (8th edn). London: Unwin Paperbacks.

Seers, D. 1959. An Approach to the Short-Period Analysis of Primary-Producing Economies. *Oxford Economic Papers*. 11 (1) February: 1–36.

Seers, D. 1962. A Theory of Inflation and Growth in Underdeveloped Economies Based on the Experience of Latin America. *Oxford Economic Papers*. 14 (2) June: 173–95.

Seers, D. 1969. The Meaning of Development. *International Development Review*. 11: 2–6.

Sen, A. 1976. Poverty: An Ordinal Approach to Measurement. *Econometrica*. 44(2): 219–31.

Sen, A. 1979. Personal Utilities and Public Judgements; or What's Wrong with Welfare Economics? *The Economics Journal*. 89: 537–58.

Sen, A. 1988. The Concept of Development. In Chenery, H. and Srinivasan, T. (eds) *The Handbook of Development Economics* Vol. 1. Amsterdam: Elsevier.

Sen, A. 1992. *Inequality Re-examined*. Oxford: Clarendon Press.

Sen, A. 1993. Capability and Well-being. In Nussbaum, M. and Sen, A. (eds) *The Quality of Life*. Oxford: Clarendon Press, 30–53.

Sen, A. 1999. *Development as Freedom*. Oxford: Oxford University Press.

Sen, S. 2005. International Trade Theory and Policy: What is Left of the Free Trade Paradigm? *Development and Change*. 36 (6) November: 1011–29.

Shafaeddin, S. 2002a. Some Implications of Accession to WTO for China's Economy. *International Journal of Development Issues*. 1 (2): 93–128. Also No. 80 in the UNCTAD Reprint Series. Accessed 28 April 2009 from: www.unctad.org.

Shafaeddin, S. 2002b. The Impact of China's Accession to WTO on the Exports of Developing Countries. UNCTAD Discussion Paper No. 160, June. Accessed 28 April 2009 from: www.unctad.org.

Shankland, A. 2000. Analysing Policy for Sustainable Livelihoods. IDS Research Report 49, September. Brighton: Institute of Development Studies.

Shaw, T. 2004. International Development Studies in the Era of Globalization and Unilateralism. *Canadian Journal of Development Studies*. XXV (1): 17–24.

Shorrocks, A. 1983. Inequality Decomposition by Factor Components. *Econometrica*. 50: 193–212.

Shorrocks, A. and van der Hoeven, R. 2004. *Growth, Inequality and Poverty: Prospects for Pro-Poor Economic Development*. London and New York: Oxford University Press.

Singer, H. 1950. The Distribution of Gains Between Investing and Borrowing Countries. *American Economic Review – Papers and Proceedings*. 40 (2 – Supplement) May: 473–85.

Smith, A. 1974. *An Inquiry into the Nature and Causes of the Wealth of Nations*. Harmondsworth: Penguin Books (originally published 1776).

Son, H. and Kakwani, N. 2003. *Poverty Reduction: Do Initial Conditions Matter?* Mimeograph. Washington, DC: World Bank.

Spraos, J. 1980. The Statistical Debate on the Net Barter Terms of Trade Between Primary Commodities and Manufactures. *Economic Journal*. 90 (357) March: 107–28.

Spraos, J. 1985. The Statistical Debate on the Net Barter Terms of Trade: A Response. *Economic Journal*. 95 (379) September: 789.

Stern, N. 1991. The Determinants of Growth. *Economic Journal*. 101 (404): 122–33.

Stern, N. 2006. *The Stern Review: The Economics of Climate Change*. Cambridge: Cambridge University Press.

Stewart, F. 2000. Income Distribution and Development. QEH Working Paper. Oxford: Queen Elizabeth House, Oxford University.

Stiglitz, J. 1989. On the Economic Role of the State. In Heertje, A. (ed.) *The Economic Role of the State*. Oxford: Basil Blackwell.

Stiglitz, J. 1997. The Role of Government in Economic Development. *Proceedings of the Annual World Bank Conference on Development Economics 1996*. Washington: The World Bank, 11–26.

Stiglitz, J. 1998a. More Instruments and Broader Goals: Moving Towards the Post-Washington Consensus. UNU WIDER Annual Lecture, Helsinki, 7 January.

Stiglitz, J. 1998b. Towards a New Paradigm for Development: Strategies, Policies and Processes. Prebisch Lecture at UNCTAD, Geneva, 19 October.

Stiglitz, J. 2002. *Globalisation and its Discontents*. London: Allen Lane.

Stiglitz, J. 2004. The Post Washington Consensus. Initiative for Policy Dialogue Working Paper. Columbia, MA: Columbia University.

Stiglitz, J. 2007. *Making Globalization Work*. Harmondsworth: Penguin Books.

Streeten, P. 1959. Unbalanced Growth. *Oxford Economic Papers*. 11 (2): 167–90.

Streeten, P. 1984. Basic Needs: Some Unsettled Questions. *World Development*. 12(9): 973–80.

Sumner, A. and Tiwari, M. 2009a. *After 2015: International Development Policy at a Crossroads*. Houndmills, Basingstoke: Palgrave Macmillan.

Sumner, A. and Tiwari, M. 2009b. After 2005: What are the Ingredients of an MDG Plus Agenda for Poverty Reduction? *Journal of International Development*. 21 (6): 534–43.

Sumner, A. and Tribe, M. 2008a. *International Development Studies: Theory and Methods in Research and Practice*. London: Sage.

Sumner, A. and Tribe, M. 2008b. Development Studies and Cross-disciplinarity: Research at the Social Science-Physical Science Interface. *Journal of International Development*. 20 (6) August: 751–67.

Sutcliffe, R. 1971. *Industry and Underdevelopment*. London: Addison-Wesley.

Sutcliffe, B. 2007. A Converging or Diverging World? In Jomo, K.S. and Baudot, J. (eds) *Flat World, Big Gaps*. London and New York: Zed Books, 48–73.

Sylvester, C. 1999. Development Studies and Postcolonial Studies: Disparate Tales of the 'Third World'. *Third World Quarterly*. 20 (4): 703–21.

Syrquin, M. 1986. Productivity, Growth and Factor Reallocation. In Chenery, H., Robinson, S. and Syrquin, M. (eds) *Industrialization and Growth: A Comparative Study*. New York: Oxford University Press, 229–62.

Tarp, F. 1993. *Stabilization and Structural Adjustment: Macroeconomic Frameworks for Analysing the Crisis in sub-Saharan Africa*. London: Routledge.

Tarp, F. and Hjertholm, P. (eds) 2000. *Foreign Aid and Development: Lessons Learnt and Directions for the Future*. London: Routledge.

Taylor, L. 1988. *Varieties of Stabilization Experience: Towards Sensible Macroeconomics in the Third World*. Oxford: Clarendon Press, Oxford University Press.

Thirlwall, A.P. 1972. *Growth and Development: With Special Reference to Developing Countries* (1st edn). London: Macmillan.

Thirlwall, A.P. 1983. Foreign Trade Elasticities in Centre-Periphery Models of Growth and Development. *Banca Nazionale del Lavoro Quarterly Review*. No. 37 September.

Thirlwall, A.P. 2000. Trade Agreements, Trade Liberalization and Economic Growth: A Selective Survey. *African Development Review*. 12 (2) December: 129–60.

Thirlwall, A.P. 2003a. *Growth and Development with Special Reference to Developing Countries* (7th edn). Houndmills, Basingstoke: Palgrave Macmillan.

Thirlwall, A.P. 2003b. *Trade, the Balance of Payments and Exchange Rate Policy in Developing Countries*. Cheltenham: Edward Elgar.

Thirlwall, A.P. 2006. *Growth and Development* (8th edn). Houndmills, Basingstoke: Palgrave Macmillan.

Thorbecke, E. and Nissanke, M. 2006. Introduction: The Impact of Globalization on the World's Poor. *World Development*. 34 (8) August: 1333–7.

Timmer, P. 1997. How Well Did the Poor Connect to the Growth Process? Consulting Assistance on Economic Reform II Discussion Paper Number 17. Harvard, MA: Harvard Institute of International Development.

Todaro, M.P. 1977. *Economic Development in the Third World* (1st edn). London: Longman.

Todaro, M. and Smith, S. 2006. *Economic Development* (9th edn). Harlow: Pearson Education.

Todaro, M. and Smith, S. 2008. *Economic Development* (10th edn). London: Pearson Addison Wesley.

Toye, J.F.J. 1987. *Dilemmas of Development: Reflections on the Counter-revolution in Development Economics*. Oxford: Blackwell.

Toye, J.F.J. 1993. *Dilemmas of Development: Reflections on the Counter-revolution in Development Economics* (2nd edn). Oxford: Blackwell.

Toye, J.F.J. 2003. Changing Perspectives in Development Economics. In Chang, H.-J. (ed.) *Rethinking Development Economics*. London: Anthem Press, 21–40.

Tribe, K. 2006. Reading Trade in the 'Wealth of Nations'. *History of European Ideas*. 32 (1) March: 58–79.

Tribe, M. 1996. Environmental Control and Industrial Projects in Less Developed Countries. *Project Appraisal*. 11 (1) March: 15–25.

Tribe, M. 1997. Sustainable Rural Economic Development: Some Theoretical and Policy Perspectives. Bureau of Integrated Rural

Development, *Proceedings of the International Seminar on Sustainable Rural Development in Sub-Saharan Africa*, 21–25 July 1997. Kumasi: University of Science and Technology, 119–28.

Tribe, M. 2000. The Concept of 'Infant Industry' in a sub-Saharan African Context. In Jalilian, H., Tribe, M. and Weiss, J. (eds.) *Industrial Development and Policy in Africa*. Edward Elgar: Cheltenham, 30–57.

Tribe, M. 2002. An Overview of Manufacturing Development in sub-Saharan Africa. In Belshaw, D. and Livingstone, I. (eds) *Renewing Development in sub-Saharan Africa: Policy, Performance and Prospects*. London: Routledge, 263–84.

Tribe, M. 2005. Industrialization and Trade Policy. In Forsyth, T. (ed.) *Encyclopedia of International Development*. Routledge: London, 350–2.

Tribe, M. 2006. Globalization, Free Trade and Market Asymmetry. In Carling, A. (ed.) *Globalization and Identity: Development and Integration in a Changing World*. London: I.B. Tauris, 43–54.

Tribe, M. and Sumner, A. 2006. Development Economics at a Crossroads: Introduction to a Policy Arena. *Journal of International Development*. 18 (7) October: 957–66.

UN 1993. *System of National Accounts*. New York: United Nations.

UN 2007. *The UN Millennium Development Goals*. New York: United Nations. Accessed 15 June 2007 from: www.un.org/millenniumgoals/.

UNCTAD 1997. *Trade and Development Report*. Geneva: United Nations Conference on Trade and Development.

UNCTAD 1999. *Trade and Development Report*. Geneva: United Nations Conference on Trade and Development.

UNCTAD 2008. *The Least Developed Countries Report 2008: Growth, Poverty and the Terms of Development Partnership*. Geneva: United Nations Conference on Trade and Development.

UNCTAD 2009a. Homepage of the website of the United Nations Conference on Trade and Development. Accessed 28 April 2009 from: www.unctad.org.

UNCTAD 2009b. *Information Economy Report 2009: Trends and Outlook in Turbulent Times*. Geneva: United Nations Conference on Trade and Development.

UNDP Annual from 1990. *Human Development Report*. New York: Oxford University Press for the United Nations Development Programme.

UNDP 1999. *Human Development Report 1999*. New York: Oxford University Press for the United Nations Development Programme.

UNDP 2008. *Human Development Report 2007–2008 – Fighting Climate Change: Human Solidarity in a Divided World*. New York: Palgrave Macmillan for the United Nations Development Programme. Accessed 5 June 2009 from: www.hdr.undp.org.

UNDP International Poverty Centre 2006. *What is Poverty? Concepts and Measures*. Poverty in Focus, December. Brasilia: UNDP IPC.

UNDP International Poverty Centre 2007. *Analysing and Achieving Pro-Poor Growth*. Poverty in Focus, March. Brasilia: UNDP IPC.

UNFCCC 2008. Investment and Financial Flows to Address Climate Change: an Update – Technical Paper. Bonn: United Nations Framework Convention on Climate Change. Accessed 22 November 2009 from: http://unfccc.int/resource/docs/2008/.

UNMP (United Nations Millennium Project) 2005. *Investing in Development: A Practical Plan to Achieve the Millennium Development Goals*. London: Earthscan.

UNRISD 1970. *Contents and Measurement of Socioeconomic Development*. Geneva: United Nations Research Institute on Social Development.

United Nations 1994. *System of National Accounts 1993*. New York, United Nations Department of Economic and Social Affairs. Website: http://unstats.un.org/unsd/sna1993/introduction.asp.

United Nations 2009a. *The Millennium Development Goals Report 2009*. New York: United Nations. Accessed 21 July 2009 from: www.un.org/millenniumgoals/.

United Nations 2009b. *Recommendations by the Commission of Experts of the President of the General Assembly on Reforms of the International Monetary and Financial System*. 20 May. Downloaded 31 May 2009 from: www.un.org/ga/president/63/interactive/financialcrisis/Preliminary Report1210509.pdf.

Van Waeyenberge, E. 2006. From Washington to Post-Washington Consensus: Illusions of Development. In Jomo, K.S. and Fine, B. (eds) *The New Development Economics: After the Washington Consensus*. London: Zed Books, 21–45.

Vernon, R. 1966. International Investment and International Trade in the Product Cycle. *Quarterly Journal of Economics*. 80 (2) May: 190–207. Reprinted in Rosenberg, N. (ed.) 1971. *The Economics of Technological Change*. Harmondsworth: Penguin Books, 440–60.

Vernon, R. 1979. The Product Cycle Hypothesis in the New International Environment. *Oxford Bulletin of Economics and Statistics*. 41 (4) November: 255–67.

Vreeland, J.R. 2003. *The IMF and Economic Development*. Cambridge: Cambridge University Press.

Vreeland, J.R. 2007. *The International Monetary Fund: Politics of Conditional Lending*. London: Routledge.

Wade, R. 2001a. Making the World Development Report 2000: Attacking Poverty. *World Development*. 29(8): 1435–41.

Wade, R. 2001b. Is Globalisation Making World Income Distribution More Equal? London School of Economics Development Studies Institute, Working Paper Series, No. 01–01.

Wade, R. 2001c. The Rising Inequality of World Income Distribution. *Finance and Development*. 38(4) December. Accessible from: www.imf.org/external/pubs/ft/fandd/2001/12/wade.htm.

Wade, R. 2005. Globalization, Poverty and Inequality. In Ravenhill, J. (ed.) *Global Political Economy*. Oxford: Oxford University Press, 292–316.

Weber, M. 1930. *The Protestant Ethic and the Spirit of Capitalism*. London: George Allen and Unwin.

Weil, D.N. 2005. *Economic Growth*. Boston: Pearson Addison Wesley.

Weisbrot, M., Baker, D., Naiman, R. and Neta, G. 2001. Growth May be Good for the Poor – But are the IMF and World Bank Policies Good for Growth? Centre for Economic Policy and Research (CEPR) Working Paper. Washington, DC: CEPR.

White, H. 1999. Global Poverty Reduction: Are We Heading in the Right Direction? *Journal of International Development*. 11 (4): 503–19.

White, H. 2002. Combining Quantitative and Qualitative Approaches in Poverty Analysis. *World Development*. 30 (3) March: 511–22.

White, H. and Anderson, E. 2001. Growth Versus Distribution: Does the Pattern of Growth Matter? *Development Policy Review*. 19 (3): 267–89.

Wikipedia 2009. *Wikipedia*. Accessed 4 June 2009 from: en.wikipedia.org/wiki/Main_Page.

Williamson, J. 1990. What Washington Means by Policy Reform. In J. Williamson (ed.) *Latin American Adjustment: How Much Has Happened?* Washington, DC: Institute for International Economics.

Williamson, J. 1993. Democracy and the Washington Consensus. *World Development*. 21 (8): 1329–36.

Williamson, J. 1994. In Search of a Manual for Technopols. In Williamson, J. (ed.) *The Political Economy of Policy Reform*. Washington: Institute of International Economics, 9–28.

Williamson, J. 2000. What Should the World Bank Think About the Washington Consensus? *World Bank Research Observer*. 15 (2): 251–64.

Williamson, J. 2004. *The Washington Consensus as Policy Prescription for Development*. Washington: Institute for International Economics. A lecture in the series 'Practitioners of Development' delivered at the World Bank on 13 January 2004: www.iie.com/publications/papers/williamson 0204.pdf.

Williamson, J. and Milner, C. 1991. *The World Economy: A Textbook in International Economics*. Hemel Hempstead: Harvester-Wheatsheaf.

Willis, K. 2005. *Theories and Practices of Development*. London: Routledge.

Winham, G. 2005. The Evolution of the Global Trade Regime. In Ravenhill, J. (ed.) *Global Political Economy*. Oxford: Oxford University Press, 87–115.

Wood, A. 2004. Making Globalization Work for the Poor: The 2000 White Paper Reconsidered. *Journal of International Development*. 16 (7) October: 934–37.

World Bank 1996. *World Development Report 1996: From Plan to Market*. Oxford: Oxford University Press for the World Bank.

World Bank 1998a. *Public Expenditure Management Handbook*. Washington, DC: World Bank.

World Bank 1998b. *RMSM-X User's Guide*. Washington, DC: World Bank Development Data Group.

World Bank 1999. *World Development Indicators 1999*. Washington, DC: World Bank.

World Bank 2000. *World Development Report 2000/1*. Washington, DC: World Bank.

World Bank 2001a. *Strengthening Statistical Systems for Poverty Reduction Strategies*. Washington, DC: World Bank.

World Bank 2001b. *Well-being Measurement and Analysis Technical Notes*. Washington, DC: World Bank.

World Bank 2002. *Poverty Reduction Strategy Papers Sourcebook.* Washington, DC: World Bank. Accessed 16 May 2009 from: www.world bank.org/prsp.

World Bank 2005. *World Development Indicators 2005*. Washington, DC: World Bank.

World Bank 2006. *World Development Report 2006*. New York: Oxford University Press for the World Bank.

World Bank 2007a. *World Development Indicators 2007*. Washington, DC: World Bank. Available from: http://go.worldbank.org/3JU2HA60D0.

World Bank 2007b. *World Development Indicators 2007 – Backmatter*. Washington, DC: World Bank. Available from: http://go.worldbank.org/3JU2HA60D0.

World Bank 2007c. *IDA's Performance-Based Allocation System: Options for Simplifying the Formula and Reducing Volatility*. Washington, DC: International Development Association, Resource Mobilization (FRM). Accessed February 2007 from: www.worldbank.org.

World Bank 2007d. *World Development Report 2007*. New York: Oxford University Press for the World Bank.

World Bank 2008. *Water Supply and Sanitation – Pricing and Subsidies*. Accessed 10 June 2008 from: www.worldbank.org.

World Bank 2009. *Debt Pocket Brochure*. Washington: World Bank. Accessed 10 June 2009 from: www.worldbank.org/hipc. See also www.worldbank.org/economicpolicyanddebt.

Zheng, 1993. An Axiomatic Characterisation of the Watts Index. *Economic Letters*. 42: 81–6.

Index

ROUTLEDGE INTERNATIONAL HANDBOOKS

Routledge International Handbooks is an outstanding, award-winning series that provides cutting-edge overviews of classic research, current research and future trends in Social Science, Humanities and STM.

Each *Handbook*:

- is introduced and contextualised by leading figures in the field
- features specially commissioned original essays
- draws upon an international team of expert contributors
- provides a comprehensive overview of a sub-discipline.

Routledge International Handbooks aim to address new developments in the sphere, while at the same time providing an authoritative guide to theory and method, the key sub-disciplines and the primary debates of today.

If you would like more information on our on-going *Handbooks* publishing programme, please contact us.

Tel: +44 (0)20 701 76566
Email: reference@routledge.com

www.routledge.com/reference

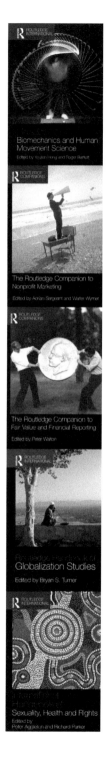

Biomechanics and Human Movement Science
Edited by Youlian Hong and Roger Bartlett

The Routledge Companion to Nonprofit Marketing
Edited by Adrian Sargeant and Walter Wymer

The Routledge Companion to Fair Value and Financial Reporting
Edited by Peter Walton

Routledge Handbook of Globalization Studies
Edited by Bryan S. Turner

Routledge Handbook of Sexuality, Health and Rights
Edited by Peter Aggleton and Richard Parker